Cambridge Elements ☰

Elements in Current Archaeological Tools and Techniques
edited by
Hans Barnard
Cotsen Institute of Archaeology
Willeke Wendrich
Polytechnic University of Turin

MACHINE LEARNING FOR ARCHAEOLOGICAL APPLICATIONS IN R

Denisse L. Argote
National Institute of Anthropology and History of Mexico

Pedro A. López-García
National Institute of Anthropology and History of Mexico

Manuel A. Torres-García
National Institute of Anthropology and History of Mexico

Michael C. Thrun
Philipps University of Marburg

COTSEN INSTITUTE OF
ARCHAEOLOGY AT UCLA

CAMBRIDGE
UNIVERSITY PRESS

![CAMBRIDGE UNIVERSITY PRESS]

Shaftesbury Road, Cambridge CB2 8EA, United Kingdom

One Liberty Plaza, 20th Floor, New York, NY 10006, USA

477 Williamstown Road, Port Melbourne, VIC 3207, Australia

314–321, 3rd Floor, Plot 3, Splendor Forum, Jasola District Centre,
New Delhi – 110025, India

103 Penang Road, #05–06/07, Visioncrest Commercial, Singapore 238467

Cambridge University Press is part of Cambridge University Press & Assessment,
a department of the University of Cambridge.

We share the University's mission to contribute to society through the pursuit of
education, learning and research at the highest international levels of excellence.

www.cambridge.org
Information on this title: www.cambridge.org/9781009506595

DOI: 10.1017/9781009506625

First published 2024

A catalogue record for this publication is available from the British Library.

ISBN 978-1-009-50659-5 Hardback
ISBN 978-1-009-50664-9 Paperback
ISSN 2632-7031 (online)
ISSN 2632-7023 (print)

Additional resources for this publication at
www.cambridge.org/argote_machine-learning

Machine Learning for Archaeological Applications in R

Elements in Current Archaeological Tools and Techniques

DOI: 10.1017/9781009506625
First published online: December 2024

Denisse L. Argote
National Institute of Anthropology and History of Mexico

Pedro A. López-García
National Institute of Anthropology and History of Mexico

Manuel A. Torres-García
National Institute of Anthropology and History of Mexico

Michael C. Thrun
Philipps University of Marburg

Author for correspondence: Denisse L. Argote, efenfi@gmail.com

Abstract: This Element highlights the employment within archaeology of classification methods developed in the field of chemometrics, artificial intelligence, and Bayesian statistics. These run in both high- and low-dimensional environments and often have better results than traditional methods. Instead of a theoretical approach, it provides examples of how to apply these methods to real data using lithic and ceramic archaeological materials as case studies. A detailed explanation of how to process data in R (The R Project for Statistical Computing), as well as the respective code, are also provided in this Element.

This Element also has a video abstract:
www.cambridge.org/EATT_Argote_MachineLearning

Keywords: archaeometry, chemometrics, artificial intelligence, Bayesian statistics, spectral analysis

ISBNs: 9781009506595 (HB), 9781009506649 (PB), 9781009506625 (OC)
ISSNs: 2632-7031 (online), 2632-7023 (print)

Contents

1 Introduction

1.1 Overview of this Volume

In recent decades, the number of archaeometric investigations that make use of physical–chemical techniques for the analysis of the composition of various archaeological materials continues to grow, as evidenced by the increasing number of publications in this area. One example of this type of studies is provenance analysis, which tries to relate archaeological materials to their original natural sources by discriminating their characteristic chemical fingerprint. In brief, it tries to determine the geological or natural origin of materials found in different archaeological contexts to establish the places of acquisition and production of the raw materials. We have chosen to approach this complex subject in two different ways, both based on very similar datasets.

In this Element, we take an applied, practical approach, allowing the reader to experiment with the provided datasets and scripts to be used in the R software package. In *Statistical Processing of Quantitative Data of Archaeological Materials,* we take a more theoretical and mathematical avenue, allowing the reader to amend and apply the discussed methods freely. These two Volumes can be used independently as well as complementary, throughout both ample cross-references are provided to facilitate the latter. As an introduction to the subject, let us first remember that the methods, basic principles and when to apply different statistical processing depends on three data scenarios: (1) when dealing with high-dimensional spectral data, (2) when employing compositional data, and (3) when managing a combination of compositional and spectral data.

Case 1 considers high-dimensionality data ($n \ll p$, where n relates to the number of observations and p are the number of variables) using full spectrum readings, such as those obtained with Fourier transform infrared spectroscopy (FT-IR), Raman spectroscopy, or X-ray fluorescence (XRF) spectroscopy. For this type of data, the suggested approach is to apply chemometric techniques and unsupervised machine learning methods. First, the spectra are preprocessed by filtering the additive and multiplicative noise, correcting misaligned peaks, and detecting outliers by robust methods. Afterwards, the data are clustered using a parametric Bayesian model that simultaneously conducts the tasks of variable selection and clustering. The variable selection employs mixture priors with a spike and slab component, which make use of the Bernoulli distributions and the Bayes factor method to quantify the importance of each variable in the clustering.

Case 2 contemplates low-dimensional data ($n > p$) where the recorded data have been converted to chemical compositions. For this case, the recommended approach is to adopt the methodology proposed by Aitchison (1986), which discusses some of the algebraic–geometric properties of the sample space of this type of data and implements log-ratio transformations. Respecting adequate preprocessing of compositional data, such as robust normalization and outlier detection, the use of model-based clustering that fits a mixture model of multivariate Gaussian components with an unknown number of components is proposed. This allows choosing the optimal number of groups as part of the selection problem for the statistical model. Mixture models have the advantage of not depending on the distance matrix used in traditional clustering analyses. Instead, the key point of the model-based clustering is that each data point is assigned to a cluster from several possible k-groups according to its posterior probabilities, thus determining the membership of each of the observations to one of the groups.

For Case 3, if reliable calibrations are available to obtain compositional data, this information can be combined with the spectra to obtain groups. For handling the data, a combination of chemometric techniques is used. In this case, a dependent variable y (or compositional values) is related to the independent variables x (or spectral values). The preprocessing is performed similarly as in Case 2; this allows calibrating a model of predictive purposes that can discriminate those variables that provide significant information to the analysis and eliminating the redundancy of information as well as collinearity. Once the selection of variables has been made, a new methodology called Databionic Swarm (DBS; implemented by Thrun, 2018) is applied for clustering the data.

To fully understand how the proposed methods work and how to apply them to your own data, these are exemplified in this Element with different case studies using quantitative data acquired from archaeological materials. The datasets used in the examples are provided in the electronic format of this Element as worksheet files with the "csv" extension. To process the data according to the exercises, the selected dataset must be imported and the source code executed in the R environment (R Development Core Team, 2011); we used version 3.6.1 on a 64-bit Windows system, although more recent versions of R are now available. R is a programming language for statistical analysis and data modeling that is used as a computational environment for the construction of predictive, classification, and clustering models. R allows you to give instructions sequentially to manipulate, process, and visualize the data. The instructions or scripts employed for each part of the process are detailed in the case studies. To learn how to employ the scripts in each step of the data processing, we encourage the reader to consult the videos associated with this Element in the electronic format of the Element.

1.2 Introduction to R

R is a public domain language and environment managed by the R Foundation for Statistical Computing (© 2016 The R Foundation) that has the virtue of being an exceptional tool for data statistical analysis and projection. This project contains a large collection of software, codes, applications, documentation, libraries, and development tools that users are free to copy, study, modify, and run. Therefore, it can be seen as a collaborative project in which anyone is invited to contribute. Although initially written by Robert Gentleman and Ross Ihaka, since 1997, it has been operated by the R Development Core Team. From early 2000 until now, it has become a kind of "standard of the scientific community." There are many publications and tutorials on its use aimed at all levels of different areas and specializations, some of which focused on the most technical and computational aspects of the language.

1.2.1 Getting Started

First, search online for CRAN R (the Comprehensive R Archive Network) or follow the link http://cran.r-project.org/ that will direct you to the web page and the instructions to download and install the latest version of R in various platforms (Linux, MacOS and Windows).

1.2.2 Data Import

Once R is installed, the next step is to call our data in the R window to be able to process them. In the screen, the indicator ">" appears and is where we must define what task we want to perform. The most commonly used configurations to perform data analysis in R are data frames, which are two-dimensional (rectangular) data structures. As in this case, we deal with datasets of low or high dimensionality, which must be arranged so that the rows in a data frame represent the cases, individuals or observations, and the columns represent the attributes or variables (see example in Table 1). These data frames must be prepared in a folder available for import and analysis.

The working directory is the place on our computer where the files we are working with are located. You can find what the working directory is with the function "> getwd ( )". You only have to write down the function and execute it. You can change the working directory using the function "setwd ( )", defining the path of the directory you want to use [e.g., `setwd("C:\obsid-ian")`]. Although there is extensive documentation on how to import/export data to R, we use the traditional method, as our data are usually in a spreadsheet with a csv extension. In R, it is sufficient to use the following command:

```
> data <- read.csv("C:\\obsidian\\mydataset.csv", header=T)
```

This command line provides the path to the folder where the data are found; the command "`read.csv`" indicates that a file with a "csv" extension is read from the "obsidian" folder located in the C root directory, and it is indicated that the data contain a name for each variable with the command "`header=T`". The symbol "`<-`" is only used for assignment; in the previous example, the file name "`mydataset.csv`" refers to the name of the data frame that you are going to work with in R. To see the structure of the data, write down the next command.

```
str(data)
```

Another way is by selecting the option "Change directory" from the File menu and navigating to the folder where our file is located. Once this path is established, you must go to the folder where your dataset is, select it and drag it with the mouse to the R console and, later, copy that path from the console and paste it into the command "`read.csv`". To see the dimensions of the dataset, you can write the function "`dim(data)`"; "`names()`" shows the names of the columns. In R, the "`summary`" function shows a general summary of the data frame variables (minimum, maximum, mean, median, first, and third quartile).

To perform an algebraic operation on a data frame such as the one exemplified in Table 1, the first column containing the identifier of each sample would have to be excluded; this is achieved by typing the following command:

Table 1 Example of a data frame of chemical concentrations of obsidian samples.

ID	Mn	Fe	Zn	Ga	Th	Rb	Sr	Y	Zr	Nb	Source
Ahuisculco	378	7468	47	18	10	109	45	20	143	19	1
Ahuisculco	379	7889	166	19	9	113	50	19	146	21	1
...	...	...	...	...	...	...	...	...	...	...	....
...	...	...	...	...	...	...	...	...	...	...	....
El Chayal	476	8162	33	17	11	146	154	20	100	9	12
El Chayal	486	8799	45	18	11	151	158	20	104	8	12
El Chayal	545	8309	36	17	10	140	151	19	101	9	12

```
data1 <- data [, 2:11]
```

In this way, only numerical variables are considered. Now you can transform all the data in the data frame, such as base 10 logarithms:

```
data2 <- log10 (data1)
```

If you also would like to transform negative values with logarithms you can use DataVisualizations::SignedLog(). Depending on the application this can be meaningful (cf, Aubert et al., 2016), even if in a strictly mathematical sense it is not allowed. If you want to see the values of a specific variable, you can do it with the following command:

```
data2$Na .
```

Some analyses require that the data be recognized as a matrix. In R, a matrix is a data structure that stores objects of the same type, conversely to a data frame, which is a rectangular array of data consisting only of numeric values. To convert a data frame to a matrix, you can use the following command:

```
newdata <- as.matrix(data2) .
```

To save a file that has been transformed, just type the following command:

```
write.csv (newdata, file="my_data.csv") .
```

This will save the file named "my_data" with a ".csv" extension to the working folder.

1.2.3 Functions

Once a dataset has been loaded, a large number of operations can be performed on it. If, for example, you have a variable "Na" from which you want to produce a histogram, it is enough to write the "hist (Na)" function to produce a bar graph of the variable. In R, the "plot ()" function is generally used to create graphs. This function always asks for an argument for the axis of the abscissa (x) and another for the ordered (y). If two variables, x and y are available, say Sr and Zr, and you want to see how these variables relate, "plot (Data2$Sr, Data2$Zr)" is enough to get the graph. The "plot ()" function allows you to customize the graph by entering titles or changing the size of the dots, the color, etc.

In addition to the "plot ()" function, there are other functions that generate specific types of graphs. In Windows, right clicking the graph can be copied to the clipboard, either as a bitmap or as a metafile. There are a variety of graphic packages for R that extend their functionality or are intended to optimize things for the user. If you have any questions about any other function, you can always access Help documentation using the "`help()`" command. For example, "`> help(mean)`" directs us to a web page where we can obtain information about the concept "mean" that corresponds to the arithmetic average.

1.2.4 Packages

R has a large number of packages that offer different statistical and graphical tools. Each package is a collection of features designed to meet a specific task. For example, there are specialized packages for data grouping, others for visualization or for data mining. These packages are hosted on CRAN [https://cran.r-project.org/]. A small set of these packets is loaded into the processor's memory when R is initialized. You can install packages using the "`install.packages()`" function, and typing in quotation marks the name of the package you want to install. They can also be installed directly from the console by going to the R menu and then selecting "Packages." For example, to install the "cluster" package, type the following:

```
install.packages("cluster").
```

After you complete the installation of a package, you can use its functions by calling the package with the "`library(cluster)`" command. Every time an operation is performed in R, it is important to use the "`rm(list = ls())`" command to delete all objects in the session and to be able to start without any remaining objects stored in the program memory. Additionally, when calling any file, it is recommended to use the "`str()`" command to know the structure of the data object.

1.2.5 Scripts

Scripts are text documents with the ".R" file extension. The scripts are the same as any text documents, but R can read and execute the code they contain. Although R allows interactive use, it is advisable to save the code used in an R file; this way, it can be used as many times as necessary. In this Element, we made use of the scripts published by the authors of the packages available for R. An advantage of these scripts and packages is that you can make use of the tutorials available for each. For example, to transform the data to the isometric log-ratio (ilr), go to CRAN [https://cran.r-project.org/] and look for the "compositions" package of van den Boogaart, Tolosana-Delgado and Bren (2023). You will find that the related script is the following:

```
## log-ratio analysis
# transformation of the data to the ilr log-ratio
library(compositions)
xxat1 <- acomp(data) # "acomp" represents one closed composition;
with this
#command the dataset is now closed
xxat2 <- ilr(xxat1) # isometric log-ratio transformation of the
data
str(xxat2)
write.csv(xxat2, file="ilr-transformation.csv")
```

In this script, the command "acomp" tells the system to consider the argument as a set of compositional values, implicitly forcing the data to close to 1. Subsequently, following Egozcue et al. (2003), the "ilr" command is used to transform the data to the isometric log-ratio, which produces compositions that are represented in Cartesian coordinates.

A more complex task can be done with R. For example, let us say that you want to implement a Principal Component Analysis with the idea of exploring the data and seeing if the first components can reveal the existence of a pattern in the data. By installing the "ggfortify" package (Horikoshi et al., 2023), it is easy to perform the analysis and graphical display of the data. For example, suppose we have a matrix of n x p with untransformed data and whose eleventh column indicates the natural source from which some obsidian samples come, such as the one in Table 1. The first step is to call the package "ggfortify" and read the data.

```
library(ggfortify)
data <- read.csv("Sources.csv", header=T) ## Sources.csv is an
example file name
str(data)
autoplot(stats::prcomp(data[-11])) ## PCA without labels; the
11th column ##is
deleted
```

In this script, the function "prcomp" will perform a principal component analysis of the data matrix and return the results as a class object; in turn, the "autoplot" function will provide the graph of the first two components. To produce a plot of the first two components that includes the provenance label for each sample and a color assignation to each group, use the following command line:

```
autoplot(stats::prcomp(data[-11]), data1 = data, colour =
'Source')
## 'Source' specifies column name keyword in your dataset
```

A biplot of the components that explain most of the variance of the data using the loading vectors and the PC scores is obtained through the following command line:

```
autoplot(stats::prcomp(data[-11]),label=TRUE,
loadings=TRUE,loadings.label=TRUE)
```

Assuming the grouping variable is available, the following command line automatically locates the centroid of each group and performs a PCA with 95 percent confidence ellipses:

```
autoplot(stats::prcomp(datos[-11]),data1=data,frame=TRUE,
frame.type='t',frame.colour=' Source')
## ` Source' specifies column name keyword in your dataset
```

In R, there are numerous clustering algorithms ranging from distance-based algorithms (e.g., to determine whether the data present a clustering structure) to more formal statistical methods based on probabilities, such as Bayesian methods or model-based clustering. For instance, for conventional clustering, the package "cluster" can be used (Maechler et al., 2022). With this package, several classical classifications can be performed by selecting both the metric used to calculate the differences between the observations and the grouping method, among which are average, single, weighted, ward, and others.

This Element provides the scripts to perform all the proposed preprocessing and clustering techniques so that the user can easily execute the commands by copying and pasting them into the R environment. For example, in Section 3, we worked with model-based clustering; for this, we employed the R libraries "Rmixmod" (Lebret et al., 2015) and "ClusVis" (Biernacki et al., 2021):

```
library (Rmixmod)
out_data<-mixmodCluster(data2,nbCluster=2:8)
summary(out_data)
plot(out_data)
library(ClusVis)
clusvisu<-clusvisMixmod(out_data)
plotDensityClusVisu(clusvisu)
```

By using the command "mixmodCluster", an unsupervised classification based on Gaussian models with a list of clusters (from two to eight clusters) is performed, determining which model best fits the data according to the BIC information criterion. In turn, the "plot()" command provides a 2D representation with isodensities, data points, and partitioning. . Alternatively, two-dimensional density-based structures can be visualized with "ScatterDensity"

(Brinkmann et al., 2023). Similarly, the "ClusVis" package (Biernacki et al., 2021) is used to obtain a graph of Gaussian components that supplies an entropic measure on the quality of the drawn overlay compared to the Gaussian clustering of the initial space. For this, only two commands are needed, "`clusvisMixmod`" and "`plotDensityClusVisu`", which are provided by the package authors.

Thus, the user has free access to the tutorials and scripts of each of the algorithms used in this Element. In many cases, the only thing that needs to be done is to replace the author's data with your own. You can also experiment with other strategies for the analysis by changing the parameters, such as the number of iterations, the initialization method, and the model selection criteria. The instructions to do so, as well as a variety of examples that the same user can reproduce, can be consulted in the documentation associated with each R package or script. It is very important to remember that the theoretical assumptions of each model must be respected; unfortunately, the data do not always conform to these. That is why it is recommended that the reader pay close attention to the theory of each method and to the behavior of his data, since a violation of the theoretical assumptions can lead to an incorrect interpretation of the data.

2 Processing Spectral Data

2.1 Applications and Case Studies

This section presents the numerical experiments conducted on archaeological materials using spectral data and the Bayesian approach. Although only examples of X-ray fluorescence data are used in this Element, the proposed methods can be applied in the same way to any other spectrometric technique such as Raman or FT-IR. To exemplify the performance of the proposed methods, two analyses were carried out, one with 156 obsidian geological samples that served as a control test (matrix available in the supplementary material as file 'Obsidian_sources.csv') and a second one using 185 ceramic fragments of archaeological interest (matrix available in the supplementary material as file 'NaranjaTH_YAcim.csv'). For the analysis of all samples, we employed a TRACER III-SD XRF portable analyzer manufactured by Bruker Corporation, with an Rh tube at an angle of 52°, a drift silicon detector and a 7.5 μm Be detector window.

The instrument was set with a voltage of 40 kV, a current of 30 μA, and a measurement time of 200 live seconds. Only for the case of the obsidian samples was employed a factory filter composed of 6 μm Cu, 1 μm Tl and 12 μm Al. A spectrum of each sample was obtained by measuring the photon emissions in 2048 channel intervals (corresponding to the energy range of 0.019 to 40 keV of the detector resolution). Is important to remember that portable

XRF is more effective for detecting Na to U elemental concentrations, which is the range that corresponds to 1.040 (in the K-alpha layer) to 13.614 keV (in the L-alpha layer). Therefore, any peaks outside of this range supply no useful information. That is the reason why the original 2,048 channels were reduced by cutting the tails of the spectra that corresponded to noninformative regions, such as low Z elements (lower than Na), the Compton peak, and the palladium and rhodium peaks (K-alpha and K-beta). For example, in the obsidian exercise, the data that were not in the range from channel 38 to channel 900 were manually deleted, leaving only the central 862 channels that correspond to the energy range of 0.74 to 17.57 keV of the detector resolution (Figure 1).

Using the numerical values obtained from the photon counts in each energy interval or channel, two n x p matrices were constructed (where n refers to the samples and p to the channel count interval):

1. An $n = 149$ x $p = 862$ matrix for the obsidians (available in the supplementary material as file 'Obsidian_Sources_38_900.csv').
2. An $n = 185$ x $p = 791$ matrix for the pottery fragments (available in the supplementary material as file 'NaranjaTH_YAcim40_830.csv').

The spectral intensities (photon counts) sampled at the given intervals (channels) represent the quantitative data employed in the statistical analysis instead of the major and trace element concentrations traditionally used for this purpose. Both datasets were preprocessed the same way. First, the EMSC algorithm, to filter the dispersion effects, and the smoothing procedure with the Savitzky–Golay algorithm were applied. With this methodology, it is not necessary to standardize each variable before using the model-based clustering since applying the EMSC filter to the data are equivalent to normalizing it. If your own spectra show any peak displacement, you should apply at this point the CluPa algorithm for peak alignment. Before developing the classification model, the atypical values were detected, removing from the matrix the samples that recorded high values in their orthogonal and score distances with the ROBPCA algorithm. It is important to note that the clustering model presupposes that all the variable-wise centers equal zero.

2.2 Exercise 1: Obsidian Samples

As mentioned in Section 2.1, the control set consisted of 149 obsidian samples of known origin and $p = 862$ energy intervals (channels). These obsidian samples were collected from 12 Central Mexico (Figure 2). The number of samples analyzed from each source is specified in Table 2. A full description of the geological setting of the obsidian sources can be found in Argote Espino et

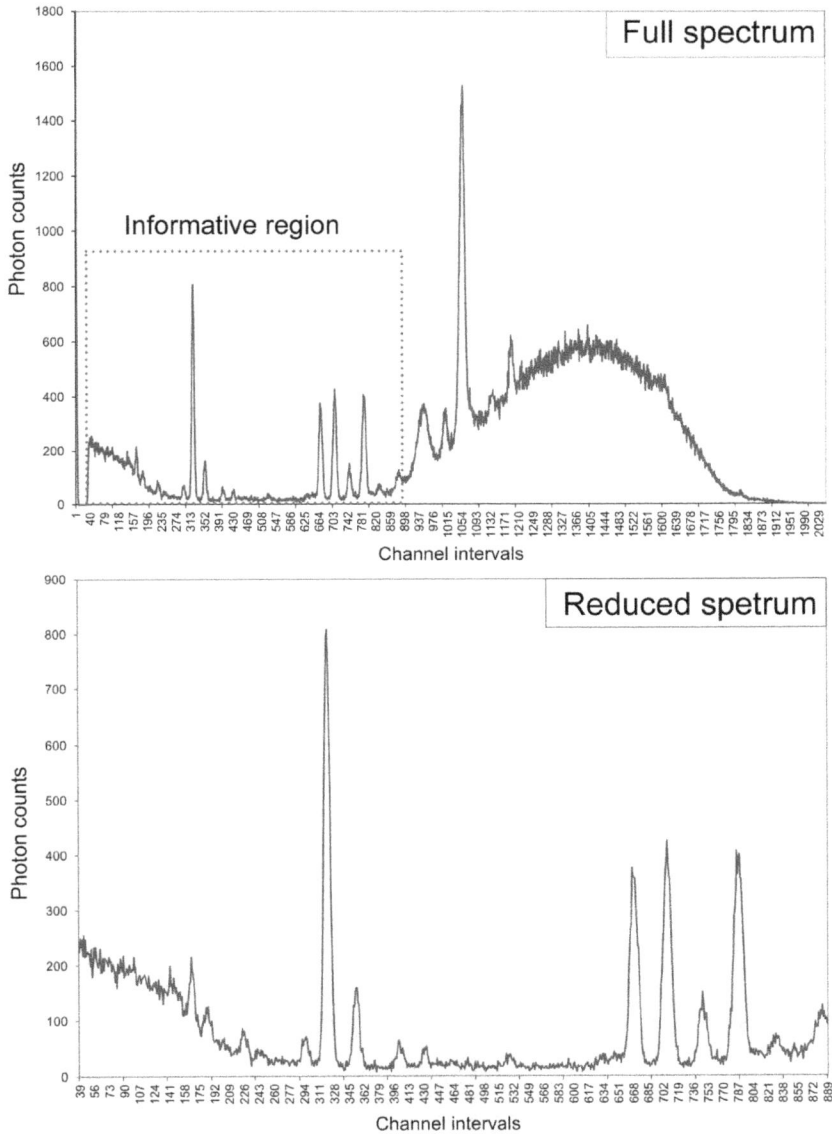

Figure 1 Comparison between the full spectrum of an obsidian sample (top) and a reduced spectrum containing only the informational region (bottom). Color version available at www.cambridge.org/argote_machine-learning

al. (2010), Argote-Espino et al. (2012), Cobean (2002), and López-García et al. (2019). These samples served as a controlled experiment for assessing the efficiency of the proposed method. The procedure is described step by step in the supplementary video "*Video 1.*"

PROCESSING

SPECTRAL DATA

Exercise 1:

Obsidian samples

Video 1 Step-by-step video on how to process spectral data of obsidian samples used in Video 1. Video files available at www.cambridge.org/argote_machine-learning

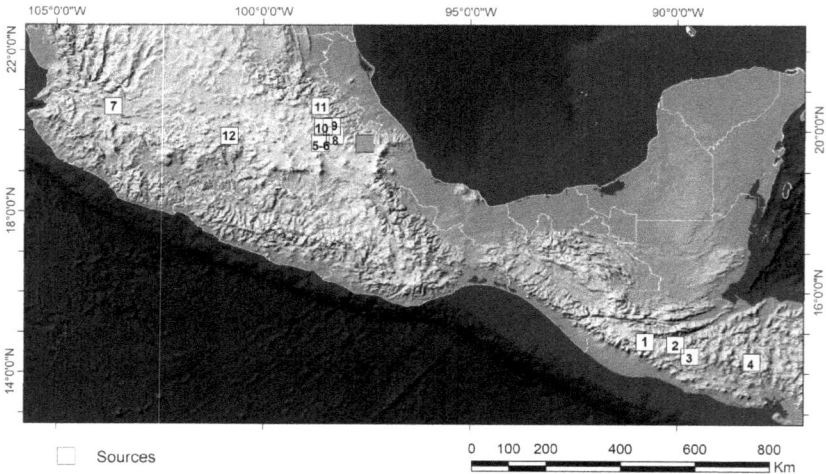

Figure 2 Geographical location of the obsidian deposits. The numbers correspond to the following sources: (1) El Chayal, (2) Ixtepeque, and (3) San Martín Jilotepeque in Guatemala; (4) La Esperanza in Honduras; and (5–6) Otumba volcanic complex, Edo. México; (7) Ahuisculco, Jalisco; (8) El Paredón, Puebla; (9) El Pizarrín-Tulancingo, Hidalgo; (10) Sierra de Pachuca, Hidalgo; (11) Zacualtipán, Hidalgo; (12) Zinapécuaro, Michoacán. Color version available at www.cambridge.org/argote_machine-learning

Table 2 Number of obsidian samples *per* location.

	Source name	Sample ID	*n*
1	El Chayal (Guatemala)	1–17	17
2	La Esperanza (Honduras)	18–33	16
3	Ixtepeque (Guatemala)	34–50	17
4	San Martin Jilotepeque (Guatemala)	51–67	17
5	Ahuisculco (Jalisco)	68–76	9
6	Otumba (Soltepec)	77–86	10
7	Otumba (Ixtepec-Pacheco-Malpaís)	87–110	24
8	El Paredon (Puebla)	111–117	7
9	El Pizarrin-Tulancingo (Hidalgo)	118–122	5
10	Sierra de Pachuca (Hidalgo)	123–132	10
11	Zacualtipán (Hidalgo)	133–142	10
12	Zinapecuaro (Michoacán)	143–149	7
		Total:	**149**

The reduced obsidian sample spectra were filtered using a combination of the EMSC + Savitzky-Golay filters; the script to perform this task is presented below.

```
## Script to filter with the EMSC algorithm version 0.9.2 (Liland
and Indahl, 2020)
rm(list = ls())
library(EMSC) #Package EMSC (Performs model-based background
correction and
# normalisation of the spectra)
dat <- read.csv("Obsidian_source_38_900.csv", header=T) # To
call the spectral data file
str(dat) # to see the data structure
dat1 <- dat[2:863] # To eliminate the first column related to the
sample identifier
str(dat1)
Obsidian.poly6 <- EMSC(dat1, degree = 6) #Filters the spectra
with a 6th order
#polynomial
str(Obsidian.poly6)
write.csv(Obsidian.poly6$corrected, file="Obsidian_EMSC.csv")
# to save the data file
# filtered with the EMSC. The User can choose other file names
#To filter the spectra with the SG filter, use the 'prospectr' pack-
age (Stevens and
```

```
#Ramirez-Lopez (2015)
library(prospectr) ## Miscellaneous functions for processing and
sample selection of
## spectroscopic data (Stevens et al., 2022)
dat2 <- read.csv("Obsidian_EMSC.csv", header=T) # Calls the file
with EMSC filtered
#data
str(dat2)
sg <- savitzkyGolay(Obsidian.poly6$corrected, p = 3, w = 11, m = 0)
write.csv(sg, file="Obsidian_EMSC_SG.csv") # The user can choose
another file name
```

Figure 3 shows the result of the filtered spectra compared to the untransformed raw data. Notice that the information was not altered. It is important to note that because we determined a polynomial of the sixth order for the SG filter, the initial matrix with $p = 862$ was reduced to $p = 852$, eliminating five channel intervals from each extreme of the data matrix.

Because the original spectra were not shifted or the intensity peaks were misaligned, it was not necessary to apply the CluPa algorithm. Nevertheless, if anyone finds it necessary, the peaks can be aligned with the following script (published by López-García et al., 2019):

```
# Run the whole script at one time
devtools::install_github("Beirnaert/speaq") # download latest
"speaq" package once!
library(speaq)
# Change file folder
your_file_path = "/Users/"
# Get the data (spectra in matrix format)
matrix3 = read.csv2(file.path(your_file_path, "your_file.csv"),
header = F, sep = ",", colClasses = "numeric", dec = ".")
spectra.matrix = as.matrix(matrix3)
index.vector = seq(1, ncol(spectra.matrix))
# Plot the spectra

speaq::drawSpec(X = spectra.matrix, main = 'mexico', xlab =
"index")
# Peak detection

peaks <- speaq::getWaveletPeaks(Y.spec = spectra.matrix,
    X.ppm = index.vector,
    nCPU = 1,
    raw_peakheight = TRUE)
# Grouping
```

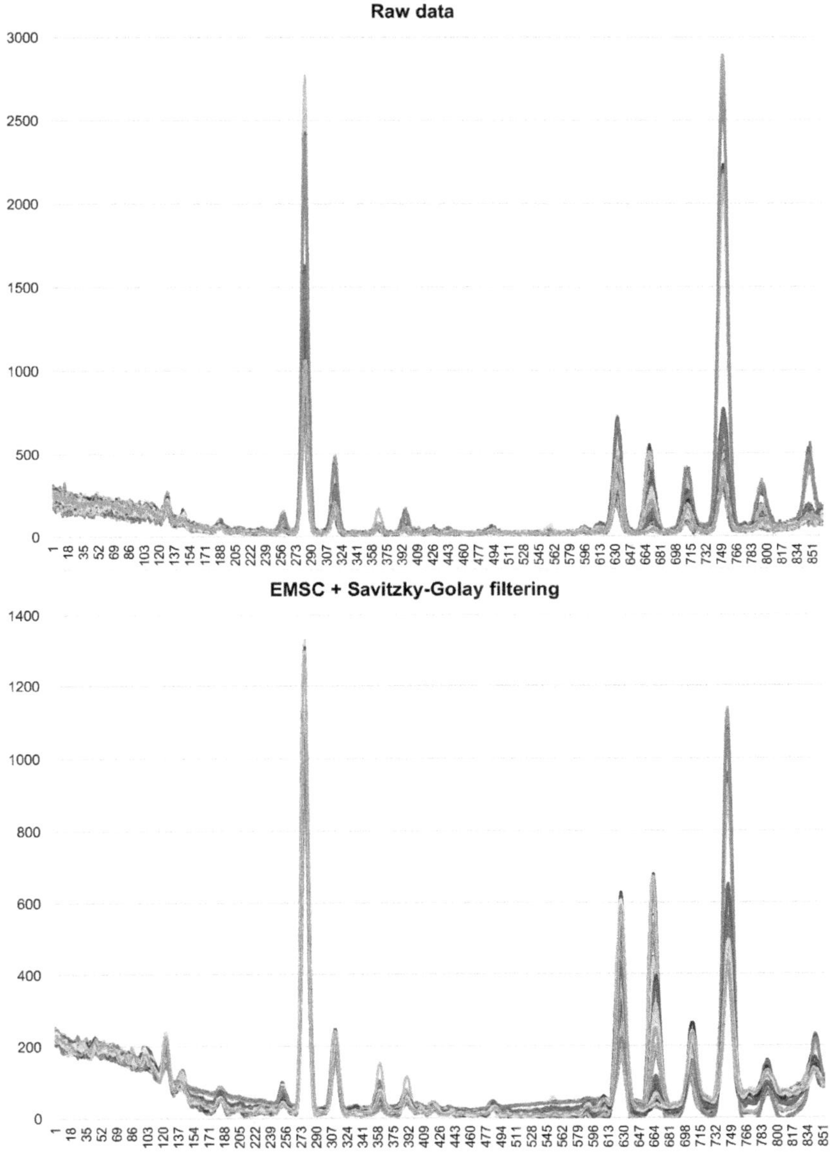

Raw data

EMSC + Savitzky-Golay filtering

Figure 3 Raw data (above) and EMSC + Savitzky–Golay filtered spectra (below). Color version available at www.cambridge.org/argote_machine-learning

```
groups<- PeakGrouper(Y.peaks = peaks)
# If the peaks are well formed and the peak detection threshold is
set low, the filling step
```

```
    #is not necessary and can be omitted
peakfill <- PeakFilling(groups_rawIntensity,
    spectra.matrix,
    max.index.shift = 5,
    window.width = "small",
    nCPU = 1)
Features <- BuildFeatureMatrix(Y.data = peakfill,
    var = "peakValue",
    impute = "zero",
    delete.below.threshold = FALSE,
    baselineThresh = 1,
    snrThres = 0)

# Aligning the raw spectra
peakList = list()
for(s in 1:length(unique(peaks$Sample))){
peakList[[s]] = peaks$peakIndex[peaks$Sample == s]
}
resFindRef<- findRef(peakList);

refInd <- resFindRef$refInd;
Aligned.spectra <- dohCluster(spectra.matrix,
    peakList = peakList,
    refInd = refInd,
    maxShift = 5,
    acceptLostPeak = TRUE,
    verbose=TRUE);
drawSpec(Aligned.spectra)
write.csv(Aligned.spectra, file =" aligned.csv")
```

After preprocessing the spectra, it is important to diagnose the data and detect outliers. For this task, use the following script extracted from the "rrcov" package (Todorov, 2020):

```
## Script to diagnose outliers
rm(list=ls())
library(rrcov)
dat <- read.csv(File_X, header=T)
str(dat)
pca <- PcaHubert(dat, alpha = 0.90, mcd = FALSE, scale = FALSE)
pca
print(pca, print.x=TRUE)
plot(pca)
summary(pca)
```

The results can be observed in Figure 4. Hubert et al. (2005) define this figure as a diagnostic plot based on the ROBPCA algorithm; it allows distinguishing regular observations and different types of outliers under the assumption that the relevant information is stored in the variance of the data (López-García et al., 2020; Thrun et al., 2023). In Figure 4, a group of orthogonal outliers (located at the top left quadrant of the graphic) that correspond to El Pizarrin-Tulancingo and Zinapecuaro source samples can be discriminated. According to this figure, there are ten observations with distances beyond the threshold of X^2 that can be considered bad leverage points or outliers; the rest are regular observations. The bad leverage points correspond to the samples from Sierra de Pachuca, which have contrasting higher chemical concentrations of Zr, Zn, and Fe than the rest of the sources (Argote-Espino et al., 2010). Therefore, they cannot be considered properly as outliers, but observations with a different behavior should not be deleted.

Once the earlier steps were concluded, we can now classify the samples. In the Bayesian paradigm, the allocation of the samples in a cluster is regarded as a statistical parameter (Partovi Nia and Davison, 2012). In general, it is better to work with the Gaussian distribution and set the default values of the hyperparameters given by the "bclust" algorithm. For this step, use the following script for R:

```
## bclust algorithm
rm(list = ls())
library(bclust) # Partovi Nia and. Davison (2015)
datx <- read.csv(File_X, header=T)
str(datx)
dat2x <- as.matrix(datx)
str(dat2x)
Obsidian.bclust<-bclust(x=dat2x,
transformed.par=c(-1.84,-0.99,1.63,0.08,-0.16,-1.68))
par(mfrow=c(2,1))
plot(as.dendrogram(Obsidian.bclust))
abline(h=Obsidian.bclust$cut)
plot(Obsidian.bclust$clust.number,Obsidian.bclust$logpos-
terior,
xlab="Number of clusters",ylab="Log posterior",type="b")
abline(h=max(Obsidian.bclust$logposterior))
str(Obsidian.bclust)
Obsidian.bclust$optim.alloc # optimal partition of the sample
Obsidian.bclust$order
# produces teeth plot useful for demonstrating a grouping on
clustered subjects
```

Robust PCA

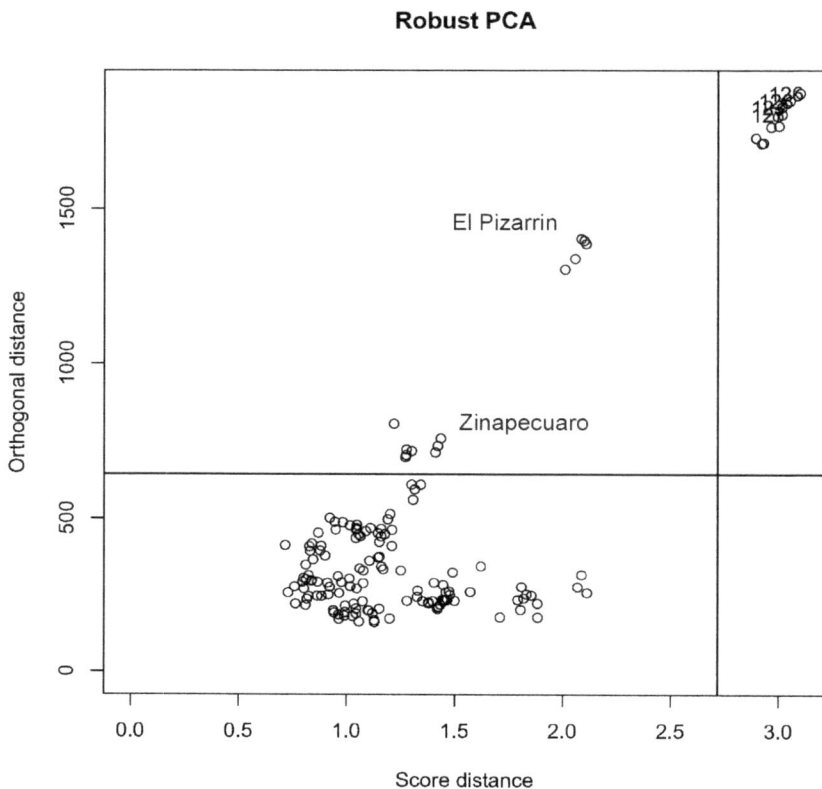

Figure 4 Diagnostic plot of the obsidian samples based on the ROBPCA algorithm. Color version available at www.cambridge.org/argote_machine-learning

```
Obsidian.bclust<-bclust(dat2x,
transformed.par=c(-1.84,-0.99,1.63,0.08,-0.16,-1.68))
dptplot(Obsidian.bclust,scale=10,varimp=imp(Obsidian.
bclust)$var,
horizbar.plot=TRUE,plot.width=5,horizbar.size=0.2,ylab.
mar=4)
#unreplicated clustering
wildtype<-rep(1,55) #initiate a vector
wildtype[c(1:3,48:51,40:43)]<-2 #associate 2 to wildtypes
dptplot(Obsidian.bclust,scale=10,varimp=imp(Obsidian.
bclust)$var,
horizbar.plot=TRUE,plot.width=5,horizbar.size=0.2,vert-
bar=wildtype,
vertbar.col=c("white","violet"),ylab.mar=4)
```

Table 3 Partition of the sample space

[1] 1 1 1 1 1 1 1 1 1 1 1 1 1 1 1 1 1 1 2 2 2 2 2 2 2 2 2 2 2 2 2
[32] 2 2 3 3 3 3 3 3 3 3 3 3 3 3 3 3 3 3 3 3 4 4 4 4 4 4 4 4 4 4 4
[63] 4 4 4 4 4 5 5 5 5 5 5 5 5 5 6 6 6 6 6 6 6 6 6 7 7 7 7 7 7 7
[94] 7 7 7 7 7 7 7 7 7 7 7 7 7 7 7 7 8 8 8 8 8 8 8 9 9 9 9 9 10 10
[125] 10 10 10 10 10 10 10 10 11 11 11 11 11 11 11 11 11 12 12 12 12 12 12 12

The result is displayed in the dendrogram of Figure 5, as well as in the partition of the sample space of Table 3. As expected, the Bayesian method clustered the obsidian samples into twelve groups (going from bottom to top in the dendrogram), all related to their geological sources: [1] Zacualtipan (ID. 133–142), which, according to the dendrogram, is subdivided into two subsources: [2] Zinapecuaro (ID. 143–149); [3] Ahuisculco (ID. 68–76); [4] El Paredon (ID. 111–117); [5] Otumba-Ixtepec, Pacheco, and Malpais (ID. 87–110); [6] Otumba-Soltepec (ID. 77–86); [7] El Pizarrin (ID.118–122); [8] Sierra de Pachuca (ID. 123–132), [9] El Chayal (ID. 1–17), which is also subdivided into two subsources: [10] La Esperanza, Honduras (ID.18–33); [11] San Martín Jilotepec (ID. 51–67), also subdivided into two subsources; and [12] Ixtepeque (ID. 34–50).

The Bayes factor (B^{10}) can be regarded as a measure of the importance that each variable holds in the classification. To determine which oligo-elements are important, the algorithm provides a list of the potentially important variables that contribute to the grouping. The following script is used for this purpose:

```
# This function plots variable importance using a barplot
Obsidian.bclust<-bclust(dat2x,
transformed.par=c(-1.84,-0.99,1.63,0.08,-0.16,-1.68),
var.select=TRUE)
Obsidian.imp<-imp(Obsidian.bclust)
#plots the variable importance
par(mfrow=c(1,1)) #retrieve graphic defaults
mycolor<-Obsidian.imp$var
mycolor<-c()
mycolor[Obsidian.imp$var>0]<-"black"
mycolor[Obsidian.imp$var<=0]<-"white"
viplot(var=Obsidian.imp$var,xlab=Obsidian.imp$labels,
col=mycolor)
```

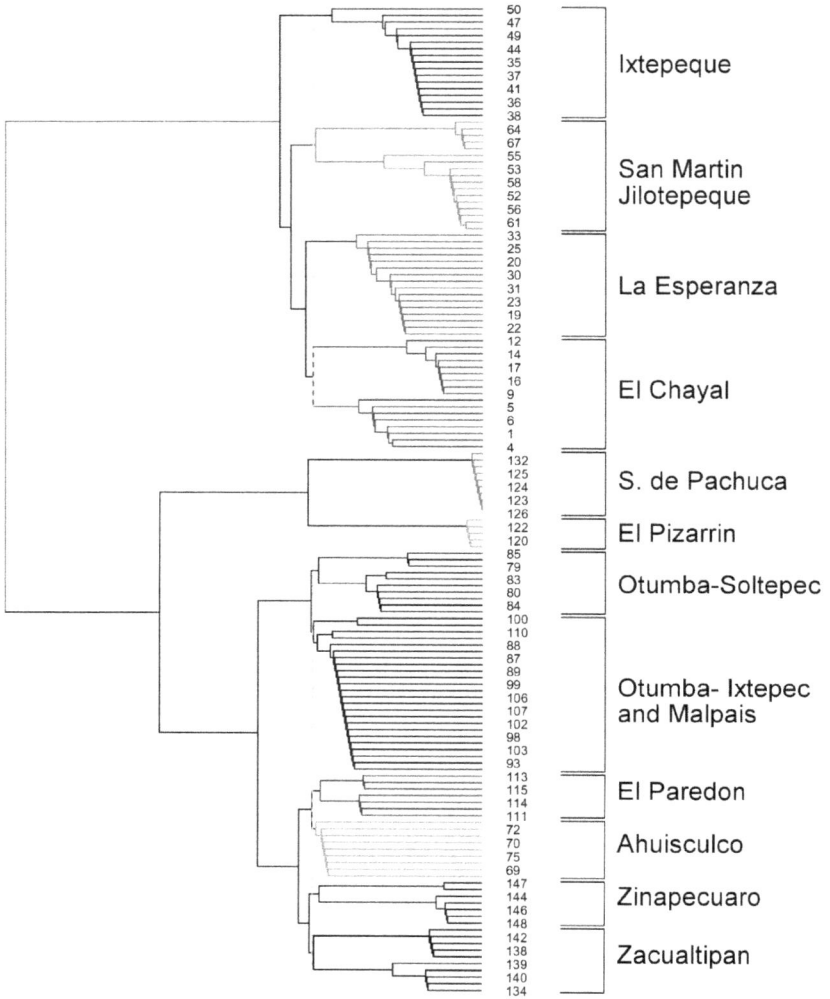

Figure 5 Dendrogram and optimal grouping found by the Gaussian model for the obsidian source data. The dendrogram visualizes the ultrametric portion of the selected distance (Murtagh, 2004). The method proposed twelve groups.
Color version available at www.cambridge.org/argote_machine-learning

```
#plots important variables in black
viplot(var=Obsidian.imp$var,xlab=Obsidian.imp$labels,
sort=TRUE,col=heat.colors(length(Obsidian.imp$var)),
xlab.mar=10,ylab.mar=4)
mtext(1, text = "Obsidian", line = 7,cex=1.5)# add X axis label
mtext(2, text = "Log Bayes Factor", line = 3,cex=1.2) #adds Y
```

```
axis labels
#Sorts the importance and uses heat colors; adds some labels to
the X and Y axes
str(Obsidian.imp)
Obsidian.imp$var
Obsidian.imp$order
```

The relevant variables can be separated from the nonrelevant variables by looking for the inflection point in the Gaussian variable selection model, such as the one observed in Figure 6, that is, the point in the distribution curve where the factor value of the variables stabilizes or stays more constant. For this case, the inflection point is presented at Log B^{10} > 1.57E+07; therefore, values of B^{10} > 1.57E+07 are considered relevant variables. The higher Bayes factors correspond to the positions of Zr, Nb, and Sr peaks (Figure 7), indicating the relative importance of these elements in the classification task. The rest of the elements had negligible and negative Bayes factors (B^{10} <1.57E+07) and hence were irrelevant.

Although discriminating groups of samples with a similar spectral profile is not a simple task, the results obtained with this clustering algorithm leave no doubt of its accuracy. First, the structure of the dendrogram was clear, and

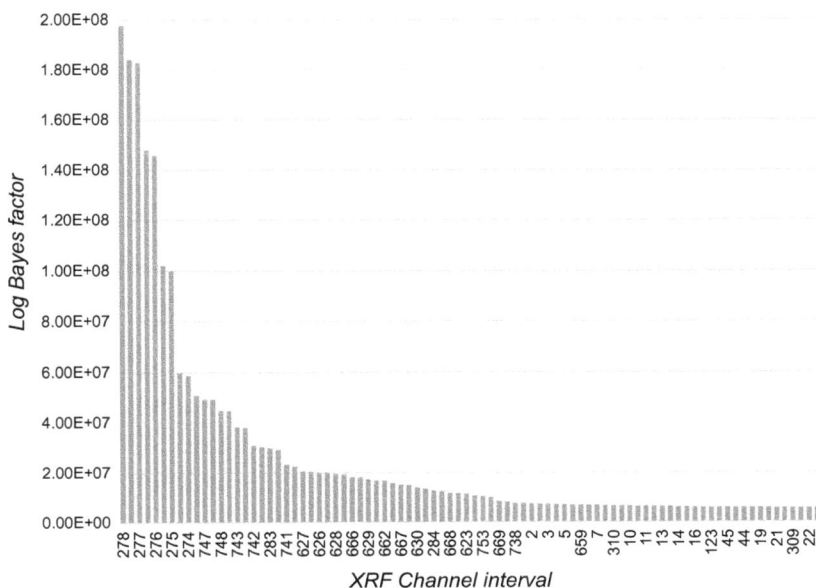

Figure 6 Log Bayes factor of variables (logB^{10}) for the Gaussian variable selection model of the obsidian data. Color version available at www.cambridge.org/argote_machine-learning

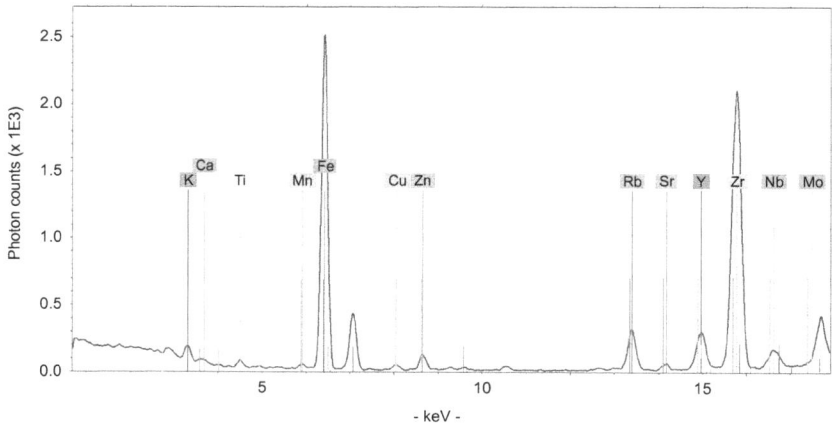

Figure 7 XRF spectrum of an obsidian sample. Color version available at www.cambridge.org/argote_machine-learning

atypical values were not observed. Second, it is possible to observe that the samples are not mixed with other groups and that each group remains characterized by its place of origin. Third, the number of groups calculated by the algorithm is correct, corresponding to the number of geological sources introduced. This allows us to conclude that each geological source has its own spectral signature, which is different from those of the other sources. It should be noted that to accurately identify the groups, it is necessary to always refer to the observations that serve as control samples (e.g., known sources).

2.3 Exercise 2: Thin Orange Pottery Samples

Thin orange ware, as its name says, is a light orange ceramic with very thin walls that became one of the main interchange products of the Classic period in Central Mexico. Its distribution over a large expanse of Mesoamerica has been considered to be closely related to the strong cultural dominion of Teotihuacan. Its wide geographical circulation has been documented in many places far from Teotihuacan (Kolb, 1973; López Luján et al., 2000; Rattray, 1979), including Western Mexico, Oaxaca, and the Mayan Highlands (i.e., Kaminaljuyú, Tikal, and Copán). The use of this ware type had a broad extension over time, starting in the Tzacualli phase (ca. 50–150 AD), peaking at the Late Tlamimilolpa and Early Xolalpan phases (350–550 AD), and declining at approximately 700 AD (Kolb, 1973; Müller, 1978).

According to Rattray (2001), the suggested chronology of the different ceramic forms of the Thin orange ware is the following. In the Tzacualli–Miccaotli phase (ca. 1–200 AD), some of the common forms are vessels with composite silhouettes, everted rims and rounded bases, vases with straight walls, pedestal base vessels, pots and a few miniatures; simple incised lines and red color decorations are present in some pieces. Some sherds found in archaeological contexts from the Miccaotli phase suggest that hemispherical forms were present. Nevertheless, the hemispherical bowls with ring bases, the most representative form in Thin orange, occur in the Early Tlamililolpan phase (ca. 200–300 AD) and continue until the end of the Metepec phase (ca. 650 AD).

Archaeological and petrographic studies performed in the 1930s (Linné, 2003) found that the components of Thin orange pastes were homogeneous and of a nonvolcanic origin. Therefore, if Teotihuacan city was settled within a volcanic region, then the production center (or at least the raw material source) should be somewhere else. These findings opened the discussion about why the most distinctive ware of Teotihuacan culture was not produced there. In the 1950s, Cook de Leonard proposed that the natural clay deposits were located south of the state of Puebla based on the material excavated from some tombs in an archaeological site near Ixcaquixtla (Brambila, 1988; Cook de Leonard, 1953).

Rattray and Harbottle performed neutron activation and petrographic analyses on samples classified as fine Thin orange ware and a coarse version of this ware called San Martin orange (or Tlajinga), the last one locally produced in Acatlán de Osorio, south of Puebla state (Rattray and Harbottle, 1992). In their conclusions, they proposed that the clay deposits and the production centers of Thin orange pottery were in the region of Río Carnero, 8 km south of Tepeji de Rodriguez town, south of Puebla state. Summarizing several former investigations about the compositional pattern of Thin orange ware, the following groups have been established:

1. A main 'Core' group, with clay and temper of homogeneous characteristics. Rattray and Harbottle (1992) and López Luján et al. (2000) mentioned that its chemical profile is characterized mainly by high concentrations of Rb, Cs, Th, and K. This group was acknowledged as "Core Thin orange" by Abascal (1974), "Thin Orange" by Shepard (1946), "group Alfa" by Kolb (1973), and "group A" by Sotomayor and Castillo (1963).

2. A coarser second group, used for utilitarian purposes (domestic ware), is characterized by having different percentages of the minerals present in the first group. This group corresponds to the "group Beta" (Kolb, 1973) and the "Coarse Thin orange" group (Abascal, 1974). Rattray and Harbottle (1992) assume that this group is formed by local imitations of the original Thin orange ware.

In this exercise, the purpose of this application is to determine possible differences in the manufacturing techniques of the Thin Orange pottery, providing a better understanding of the underlying production processes. The focus is on identifying natural groups with homogeneous chemical compositions within the data, leading to the determination of whether this ceramic type was crafted following a unique recipe (clay and temper) or if there were several ways to produce it. By comparing our results with those obtained by other researchers on the conformation of a single 'core' group (Abascal, 1974; Harbottle et al., 1976; Rattray and Harbottle, 1992; Shepard, 1946), new evidence could be provided that might help refine the current classification of this significant ware.

The procedure is shown step by step in the supplementary video *"Video 2"*. The archaeological pottery set consisted of 176 ceramic fragments and 9 clay samples (extracted from a natural deposit near the Rio Carnero area). Both sets of materials were analyzed with a portable X-ray fluorescence spectrometer. To conduct the comparative analysis with adequate variability, it was necessary to collect several samples of the same ceramic type from different locations and contexts. Therefore, the pottery samples were provided by different research projects that performed systematic excavations at various Central Mexico archaeological sites (Figure 8; Table 4).

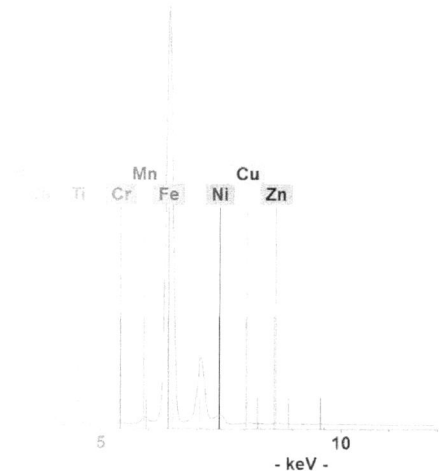

PROCESSING SPECTRAL DATA

Exercise 2:

Ceramic

samples

Video 2 Step-by-step video on how to process spectral data of ceramic samples used in Video 2. Video files available at www.cambridge.org/argote_machine-learning

Table 4 Number of Thin orange pottery samples and locations

Archaeological site	Mexican state	N
Teteles de Santo Nombre	Puebla	23
Izote, Mimiahuapan, Mapache, and other sites	Puebla	22
Huejotzingo	Puebla	6
Xalasco	Tlaxcala	53
Teotihuacan	Mexico	72
Clay source	Puebla	9
	Total:	**185**

Pottery is a very heterogeneous material; therefore, the ceramic samples were treated differently. Following the recommendations of Hunt and Speakman (2015), the ceramic samples were prepared as pressed powder pellets. From each pottery sherd, a fragment weighing approximately 2 g was cut. The external surfaces of each fragment were abraded with a tungsten carbide handheld drill, reducing the possibility of contamination from depositional processes. The residual dust was removed with pressurized air, and the fragments were pulverized in an agate mortar. After grinding and homogenizing, the powder was compacted into a 2-cm diameter pellet by a cylindrical steel plunger with a manually operated hydraulic press. No binding agent was added. These pellets provided samples that were more homogeneous and with a uniformly flat analytical surface.

Figure 9 displays an example XRF spectrum of one Thin orange pottery sample. The main elements in the ceramic matrix are iron (Fe), followed by calcium (Ca), potassium (K), silicon (Si), and titanium (Ti); aluminum (Al), manganese (Mn), nickel (Ni), rubidium (Rb), and strontium (Sr) are also present at lower intensities. Sulfur (S), chromium (Cr), copper (Cu), and zinc (Zn) can be considered trace elements.

In this case, the matrix has many more variables than observations ($n \ll p$), with $p = 2{,}048$; thus, much of the information contained was irrelevant for clustering. As mentioned at the beginning of this section, it was decided to manually cut some readings as they contained values close to zero or corresponded to undesirable effects, such as light elements below detection limits, the Compton peak, Raleigh scattering and palladium and rhodium peaks (produced by the instrument). The cuts were made at the beginning (from channel 1 to 39) and end (from channel 831 to 2,048) of the spectrum, retaining the elemental information corresponding to the analytes between channels 40 and 830, related to the energy range of 0.78 to 16.21 keV of the detector resolution. In this way,

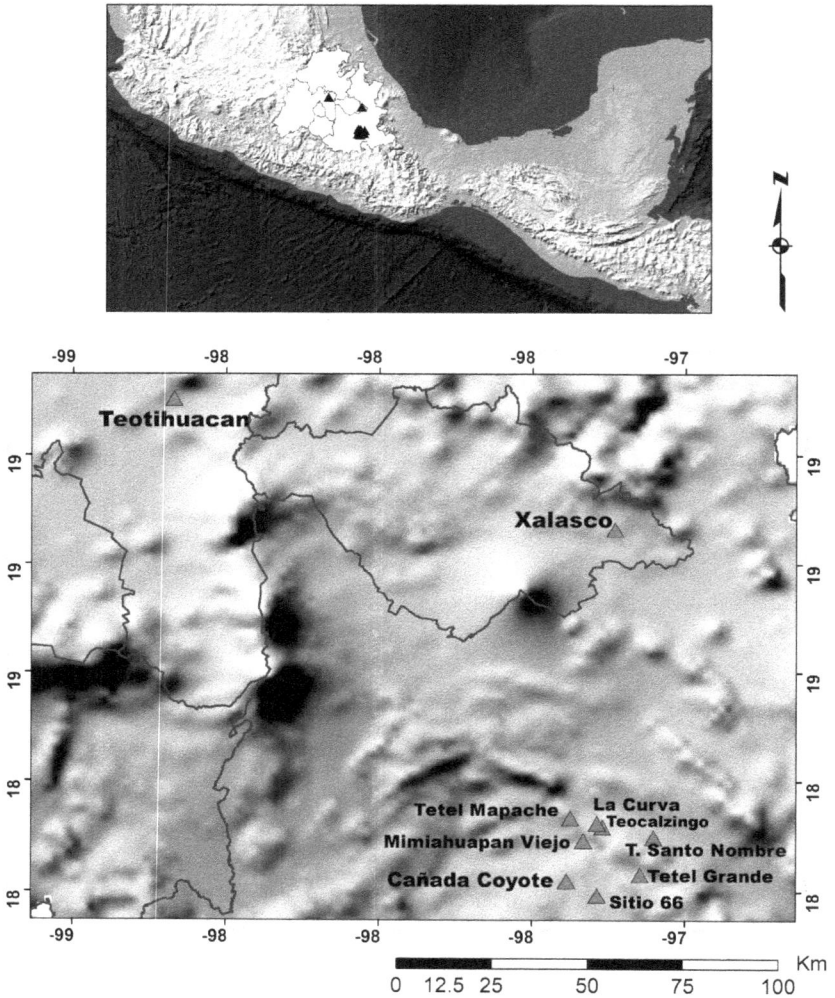

Figure 8 Geographical location of the archaeological sites from which the Thin orange pottery samples were collected. Color version available at www.cambridge.org/argote_machine-learning

only $p = 791$ channel intervals were kept. This matrix can be found in the supplementary material as file 'NaranjaTH_YAcim40_830'. It should be noted that because there were no displacements in the spectra, it was not necessary to use the peak alignment algorithm.

The next step was to filter the spectra as in the previous study case, using only the EMSC, as the spectrum did not contain scattering effects. For this purpose, use the following script:

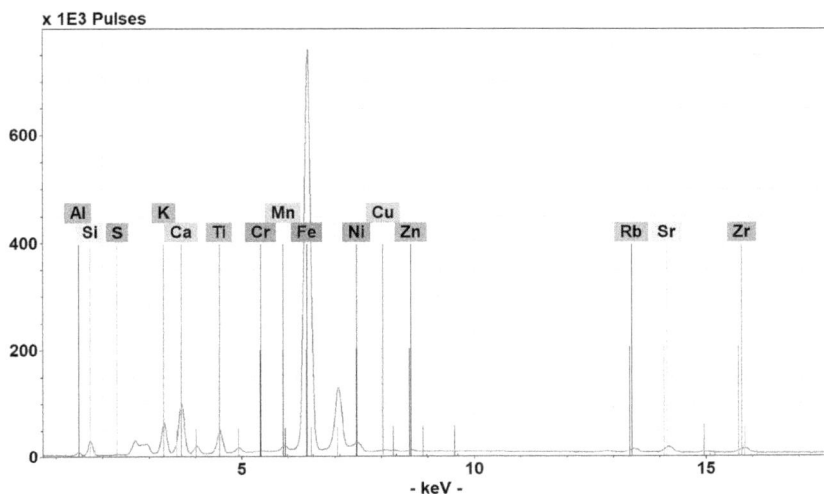

Figure 9 XRF spectrum of a representative sample of Thin orange pottery. Color version available at www.cambridge.org/argote_machine-learning

```
## Script to filter with the EMSC algorithm
rm(list = ls())
library(EMSC) #Package EMSC (Performs model-based background
correction and
# normalization of the spectra)
dat <- read.csv("C:\\NaranjaTH_YAcim40_830.csv", header=T) #
Calls the spectral data
#file
str(dat) # to see the structure of the data
dat1 <- dat[2:792] # Eliminates the first column related to the
sample identifier
str(dat1)
pottery.poly6 <- EMSC(dat1, degree = 6) #Filter spectra with a
6th-order polynomial
str(pottery.poly6)
write.csv(pottery.poly6$corrected, file="pottery_EMSC.csv")
# to save the data file
# filtered with the EMSC
```

Once filtering was performed, the diagnosis of outliers was performed with the following script:

```
## Script to diagnose outliers (Todorov, 2020)
rm(list=ls())
library(rrcov)
```

```
dat <- read.csv(File_X, header=T)
str(dat)
pca <- PcaHubert(dat, alpha = 0.90, mcd = FALSE, scale = FALSE)
pca
print(pca, print.x=TRUE)
plot(pca)
summary(pca)
```

In the outlier detection with the ROBCA algorithm (Figure 10), most of the ceramic sample data vectors have regular patterns with normal punctuation and orthogonal distances. We can distinguish seventeen orthogonal outliers in the upper left quadrant of the graph, one observation with an extreme orthogonal distance (observation no. 126), and a small group of ten bad leverage points in the upper right quadrant. The robust PCA high-breakdown method treats this last group as one set of outliers. An interesting fact about the set of detected bad

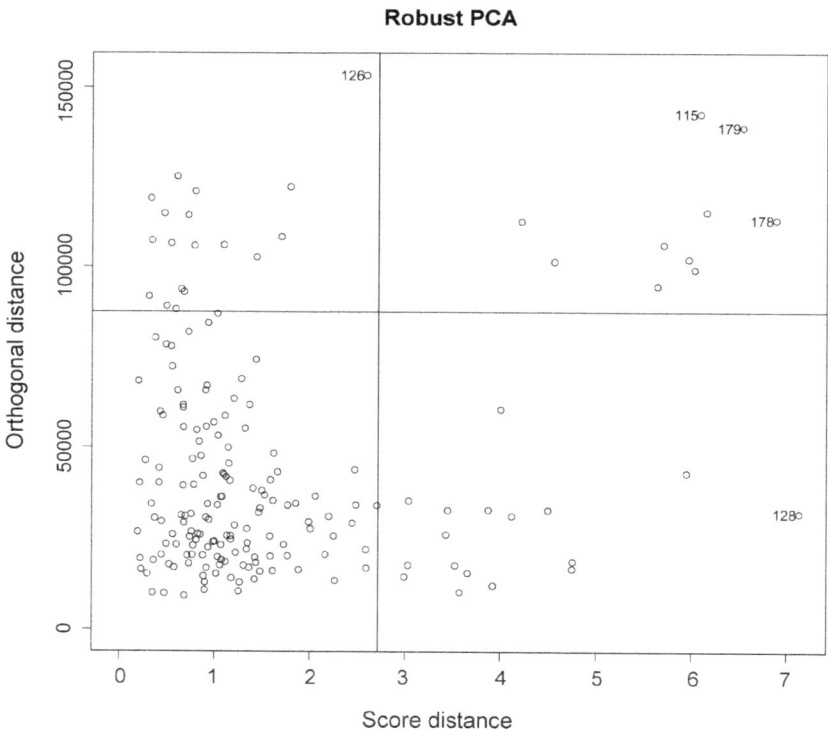

Figure 10 Outlier map of the Thin Orange pottery dataset computed with ROBPCA based on five principal components. Color version available at www.cambridge.org/argote_machine-learning

leverage points is that nine of them correspond to clay deposits (cases no. 177 to 185); only one sample (case no. 115) is related to an archaeological site in Puebla that shows a low amount of manganese content. Therefore, instead of just being measurement errors, outliers can also be seen as data points that have a different origin from regular observations, such as the case of the pure clay samples. According to this, no observations were removed from the analysis.

For the Bayesian clustering, the model parameters were set the same way as in the obsidian case. The algorithm provides a list of the potentially important variables that contribute to the clustering. In this case, the variable selection extension of the Gaussian model (Figure 11) selected 22 of the 791 initial variables as the most important ones. These twenty-two variables (channels) corresponded to the energy ranges of Fe (6.3 to 6.55 and 7 to 7.11 KeV) and Ca (3.7 to 3.75 KeV) chemical elements. Calcium and iron oxides (such as hematite) are two components that are commonly found in pottery and mudrock composition at variable concentrations depending on the parental material (Callaghan et al., 2017; Minc et al., 2016; Ruvalcaba et al, 1999; Stoner, 2016). These results are different from Rattray and Harbottle (1992) analysis in which the pottery was mainly determined by high concentrations of Rb, Cs, Th and K. On the other hand, Kolb (1973) found that Fe and Ti were important elements present in his Alpha and Beta groups.

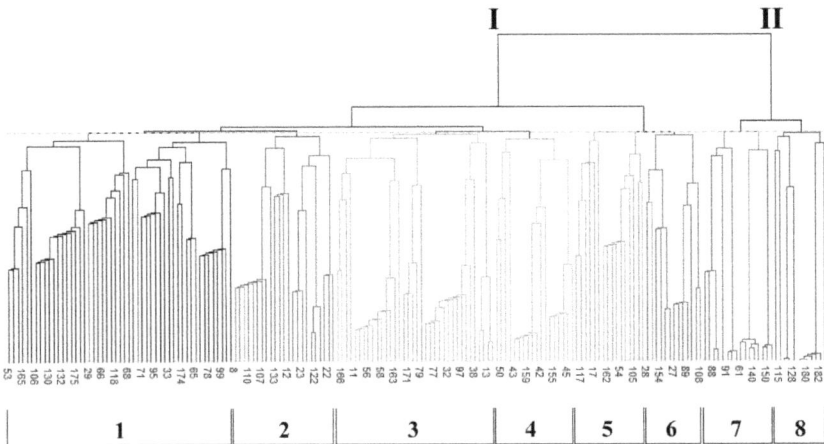

Figure 11 Log Bayes factor ($\log B^{10}$) for the Gaussian variable selection model of Thin orange data. The Bayes factors are computed for the optimal grouping found by agglomerative clustering using the Gaussian model. Color version available at www.cambridge.org/argote_machine-learning

The resulting dendrogram (Figure 12) grouped the data into two main clusters (**I** and **II**) subdivided into six (1 to 6) and two subgroups (7 and 8), respectively. Subgroups 7 and 8 showed chemical differences that distinctively separated them from the rest of the groups. The number of samples (n) assigned to each subgroup were as follows: Group 1 = 23, Group 2 = 51, Group 3 = 16, Group 4 = 36, Group 5 = 13, Group 6 = 12, Group 7 = 18, and Group 8 = 16. Samples from group 1 come mostly from the archaeological site of Xalasco. Samples from Groups 2, 3, 6, 7, and 8 come from Teotihuacan, Xalasco, and several Puebla sites. Group 4 contains samples from some sites in Puebla State and the northeastern sector of Teotihuacan city. Group 5 has samples mainly from Teteles del Santo Nombre and a few from Xalasco and Teotihuacan. Groups 4 and 6 contain the clay samples from the Rio Carnero area. Table 5 summarizes the ceramic shapes included in each group, showing a great variability of forms in each group.

The results obtained in this spectral analysis revealed the existence of two large groups subdivided into several subgroups that exhibit a certain degree of chemical differentiation, indicating that different raw materials were used to produce the Thin Orange ware. Pottery is produced by mixing clays and aplastic particles or temper, with the clay predominating over the temper. In this case, the clay deposit

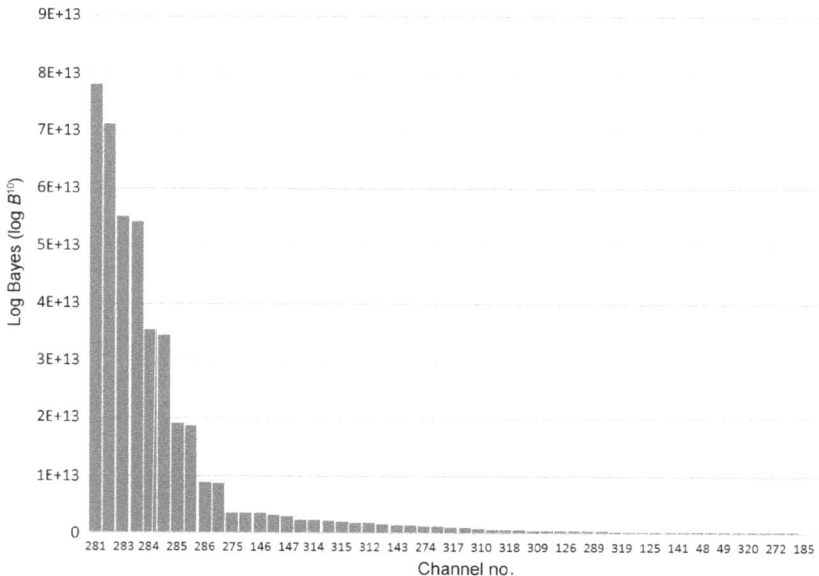

Figure 12 Dendrogram and optimal grouping found by the Gaussian model for the Thin Orange samples. The horizontal bar at the bottom refers to the optimal grouping found by the Gaussian model. Color version available at www.cambridge.org/argote_machine-learning

Table 5 Number of samples (*n*) and ceramic forms included in each subgroup:

Group 1 (*n* = 23)	Group 2 (*n* = 51)	Group 3 (*n* = 16)	Group 4 (*n* = 36)
Hemispherical bowl with ring base (13 sherds)	Hemispherical bowl with ring base (12 sherds)	Bowl with ring base (4 sherds)	Bowl with ring base (11 sherds)
Vase with incised exterior decoration (1 sherd)	Hemispherical bowl (6 sherds)	Cylindrical vase (4 sherds)	Hemispherical bowl (3 sherds)
Tripod vase with nubbin supports (1 sherd)	Vessel with convex wall (4 sherds)	Jar with incised exterior decoration (2 sherds)	Cylindrical vase (4 sherds)
Jar with incurved rim and incised simple double-line (1 sherd)	Cylindrical vase (5 sherds)	Tripod vase with nubbin supports (1 sherd)	Incense burner (1 sherd)
Jar with incised exterior decoration (1 sherd)	Tripod vase with nubbin supports (2 sherds)	Undetermined shape (5 sherds)	Basin (1 sherd)
Pot (1 sherd)	Vase with small appliqués (1 sherd)		Vessel with pedestal base (1 sherd)
Undetermined shape (5 sherds)	Tripod vessel with deep parallel grooves (1 sherd)		Vessel with incised exterior decoration (1 sherd)
	Vessel with pedestal base (1 sherd)		Tzacualli phase Pot (1 sherd)
	Miniature vessel (1 sherd)		
	Undetermined shape (18 sherds)		Undetermined shape (12 sherds)
			Natural clay deposit (1 sample)

Table 5 (cont.)

Group 5 (n = 13)	Group 6 (n = 12)	Group 7 (n = 18)	Group 8 (n = 16)
Bowl with ring base (4 sherds)	Hemispherical bowl with ring base (3 sherds)	Hemispherical bowl with ring base (7 sherds)	Hemispherical bowl with ring base (5 sherds)
Vessel with pedestal base (1 sherd)	Jar with incised exterior decoration (1 sherd)	Hemispherical vessel (2 sherds)	Annular-based hemispherical bowls and incised exterior decoration (1 sherd)
Vessel with incised exterior decoration (1 sherd)	Natural clay deposit (8 samples)	Vessel with recurved composite wall (1 sherd)	Tripod vessel with everted rim (1 sherd)
Cylindrical vase (3 sherds)		Vessel with exterior punctate decoration (1 sherd)	Vessel with red pigment and incised exterior decoration (1 sherd)
Vessel with convex wall and flat-convex base (1 sherd)		Pot (1 sherd)	Jar with incised exterior decoration (1 sherd)
Undetermined shape (3 sherds)			Undetermined shape (6 sherds)

samples were classified inside the main group **I** (in subgroups 1 and 6); thus, it can be assumed that, for manufacturing the pieces related to this main group, clay banks from the same geological region were used. The Rio Carnero region is shaped by a set of deep *barrancas* (canyons) predominantly from the Acatlán Complex, the geological vestige of a Paleozoic ocean formed between the Cambro-Ordovician and late Permian periods (Nance et al., 2006). This region contains banks of schists rich in hematite located in the ravines of Barranca Tecomaxuchitl and Rio Axamilpa. On the other hand, the chemical differences of main group **II** (with only 34 samples) indicate the extraction of clay from a different region.

According to the results, an interpretation can be as follows. The division of the samples into two main groups seems to be associated with two different clay deposit regions from which the raw material was exploited. The internal differences in their chemical composition, probably related to differences in temper, influenced the clustering algorithm to classify them into separate subgroups. This could mean that the aplastic particles used in the mixture for manufacturing the ceramic pieces did not naturally occur in the clay and were added by the artisans. The last observation is consistent with Kolb (1973), who stated that the temper was deliberately added and is not found *in situ* in the natural clay deposit.

The considerable variety of patterns presented by each of the eight subgroups suggests that the recipe for manufacturing the pieces was not used uniformly and that multiple ceramic production centers existed, employing their own and specific production recipes. In other words, each center would have produced its version of the Thin orange pottery with a standardized composition, and this was different to some extent from the ceramic made in other workshop centers. The results also support the idea that there was compositional continuity through time, despite the different shapes of the analyzed pieces.

3 Processing Compositional Data

3.1 Applications and Case Studies

In this section, data from published case studies were used to illustrate the techniques for processing compositional data. In summary, the steps for handling all datasets are as follows. First, the data are rescaled in such a way that the sum of the elemental concentrations row is equal to 100 percent. Afterwards, the data are transformed to log-ratios using the *ilr* transformation, translating the geometry of the Simplex into a real multivariate space. Once the *ilr* coordinates are obtained, data are standardized using a robust min/max-standardization. Any observation with a value equal to zero is imputed. As a diagnosis method, an MCD estimator is applied to identify the presence of outliers in the data. Then, model-based clustering is employed for the classification and visualization of the data. Finally, an SBP is calculated and graphically represented using the CoDa-dendogram for understanding differences in the composition of the clusters.

3.2 Exercise 1: Unsupervised Classification of Central Mexico Obsidian Sources

In this case study, the compositional values of obsidian samples collected from different natural sources (Table 6) located in Central Mexico, published by Lopez-Garcia et al. (2019), and Guatemala, retrieved from Carr (2015), were processed. The procedure is described step by step in the supplementary video "*Video 3*". The intention of this analysis is to demonstrate the performance of the model-based clustering and an unbiased visualization system in a controlled environment. The dataset was ideal for the analysis because the correct number of clusters and the cluster to which each of the observations belonged were known.

PROCESSING COMPOSITIONAL DATA

Exercise 1: Unsupervised Classification of

Central Mexico Obsidian Sources

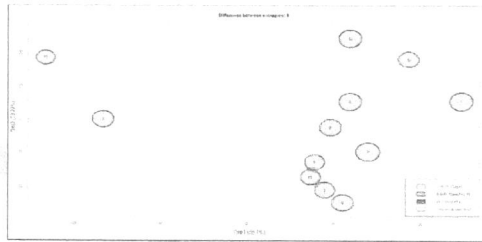

Video 3 Step-by-step video on how to process compositional data of the obsidian samples used in Video 3. Video files available at www.cambridge.org/argote_machine-learning

The dataset consisted of $n = 136$ samples with $p = 10$ variables containing the elemental composition of the samples (Mn, Fe, Zn, Ga, Th, Rb, Sr, Y, Zr, and Nb), obtained with a portable X-ray fluorescence (pXRF) instrument. This matrix is available in the supplementary material as file "Mayas_sources_onc. csv". In this example, we considered the problem of determining the structure of the data without prior knowledge of the group membership. The estimation of the parameters was performed with the maximum likelihood, and the best

Table 6 Samples collected from Mesoamerican obsidian sources.[a]

Source name	Geographic region	Sample ID no.	N
Ahuisculco	Jalisco	1–9	9
El Chayal	Guatemala	10–25	17
San Martin Jilotepeque	Guatemala	26–39	14
Ixtepeque	Guatemala	40–55	16
Otumba (Soltepec)	Edo. de México	56–65	10
Otumba (Ixtepec-Malpais)	Edo. de México	66–88	23
Oyameles	Puebla	89–95	7
Paredon	Puebla	96–102	7
Tulancingo-El Pizarrin	Hidalgo	103–107	5
Sierra de Pachuca	Hidalgo	108–117	10
Zacualtipan	Hidalgo	118–127	10
Zinapécuaro	Michoacán	128–134	8

[a] The compositional values of samples from El Chayal, San Martin Jilotepeque, and Ixtepeque were retrieved from Carr (2015).

model was selected using the ICL criterion, resulting in $K = 12$ components. Compositional data are constrained data and therefore must be translated into the appropriate geometric space. To convert the data to completely compositional data, the following code from the "compositions" package in R is used (van den Boogaart et al., 2023):

```
rm(list=ls()) # Delete all objects in R session
## log-ratio analysis
# load quantitative dataset. The name of the file for this example
is Sources.csv. You can
#change it for the location and name of your own data file.
data <- read.csv("C:\\obsidian\\Mayas_sources_onc.csv",
header=T)
str(data) # displays the internal structure of the file, including
the format of each column
dat2 <- data[2:11] # delete data identification column
str(dat2)
# transformation of the data to the ilr log-ratio
library(compositions) ## van den Boogaart, Tolosana-Delgado and
Bren (2023)
xxat1 <- acomp(dat2) # the function "acomp" representing one
closed composition.
```

```
# With this command, the dataset is now closed.
xxat2 <- ilr(xxat1) # isometric log-ratio transformation of the
data
str(xxat2)
write.csv(xxat2, file="ilt-transformation.csv") #You can
choose a personalized file name
```

In the isometric transformation output file, the geometric space is $D - 1$. Once the data have been taken to the Simplex geometry, it is important to normalize them robustly so that there are no variables with greater weight. This is achieved by loading the *clusterSim* package in R with the normalization option = na3 using the code presented below:

```
# Robust normalization
# In our case, the file "xxat2" that contains the isometric
transformation of the #compositional data was normalized with
the robust equation presented in the #transformations section.
library(clusterSim) ## Walesiak and Dudek (2020)
z11<- data.Normalization(xxat2,type="n3a",normalization=
"column",na.rm=FALSE)
# This corresponds to the robust normalization described in
'Compositional and
#Completely compositional data' Section of Volume I.
# n3a positional unitization ((x-median)/min(x) - max(x))
z12 <- data.frame(z11) # After the previous operations, it is
necessary to convert the
#data output to a data frame to tell the program that the
observations are in the
#rows and columns represent the attributes (variables)
str(z12)
```

After transforming the data through robust normalization, it is convenient to verify that there are no values equal to 0. In this example, no zero values were present; thus, no imputation was needed. In the case that your data have values equal to zero, it is recommended to employ an imputation algorithm such as Amelia II:

```
# Imputation of data
# Loads the user interface to perform the imputation of values
Library (Amelis)
AmeliaView()
```

Afterwards, to identify the presence of outliers in the data, a diagnosis is performed through the MCD estimator of the *rrcov* package (Todorov and Filzmoser, 2009):

```
# Parameters of the model
# kmax maximal number of principal components to compute. The
default is kmax=10.
# Default k= 0; if we do not provide an explicit number of compo-
nents, the algorithm
# chooses the optimal number. alpha: 0.7500; this parameter
measures the fraction
# of outliers the algorithm should resist (default).
# The matrix dimension in this example is n = 136 and p = 10
library(rrcov) ## Todorov (2020)
MCD_1D <- data.matrix(z12[, 1:9])
cv <- CovClassic(MCD_1D)
plot(cv)
rcv <- CovMest(MCD_1D)
plot(rcv)
summary(MCD_1D)
```

Figure 13 shows the distance–distance plot, which displays the robust distances *versus* the classical Mahalanobis distances (Rousseeuw and van Zomeren, 1990), allowing us to classify the observations and to identify the potential outliers. Note that the choice of appropriate distance metric is essential (Thrun, 2021a). The dotted line represents points for which the classical and robust distances are equal. Vertical and horizontal lines are plotted in values $x = y = \sqrt{X_\rho^2}, 0.975$; points beyond this threshold are considered outliers. While the robust estimation detects many observations whose robust distance is above the threshold, the Mahalanobis Distance classifies all points as regular observations. Observations with large robust distances are not candidates for outliers because they do not have an impact on the estimates. The only two observations that exceed the values of X^2 are those with ID numbers 9 and 136, which are convenient to eliminate from the estimate and leave only $n = 134$ samples in the dataset.

Once the preprocessing of the data has concluded, the mixed model of multivariate Gaussian components is adjusted for clustering purposes using the *Rmixmod* and *Clusvis* packages:

```
# Classification with Mixture Modeling: Clustering in Gaussian
case
# To fit the mixture models to the data and for the classification,
two additionaL
#programs must be loaded.
rm(list=ls())
library(Rmixmod) ## Lebret et al., 2015)
```

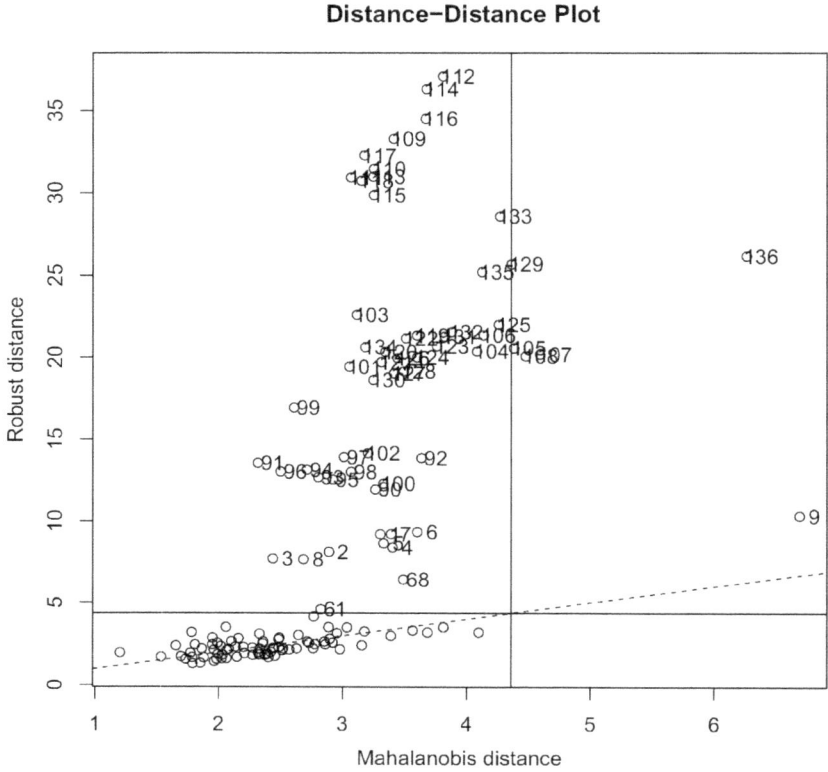

Figure 13 Distance–distance plot of the samples from obsidian sources.
Color version available at
www.cambridge.org/argote_machine-learning

```
library(ClusVis) ## (Biernacki et al., 2021)
fammodel <- mixmodGaussianModel(family="general",equal.
proportions=FALSE)
Mod1<-mixmodCluster(data,12, strategy = mixmodStrategy(algo =
"EM", nbTryInInit = 50, nbTry=25))
#EM (Expectation Maximization) algorithm
# nbTryInInit: integer defining the number of tries in the
initMethod algorithm.
# nbTry: integer defining the number of tries
summary(Mod1)
Mod1["partition"] # partition output made by mixing model
## Gaussian-Based Visualization of Gaussian and Non-Gaussian
Model-Based Clustering
library(ClusVis) ## (Biernacki et al., 2021)
resvisu <- clusvisMixmod(Mod1) # Gaussian-Based Visualization
```

```
of Gaussian Model-Based #Clustering
plotDensityClusVisu(resvisu) #probabilities of classification
are generated by using the model #parameters.
```

In this script, the command *mixmodGaussianModel* is an object defining the list of models to run. The function *mixmodCluster* computes an optimal mixture model according to the criteria supplied and the list of models defined in [Model] using the algorithm specified in the command 'strategy' [strategy = mixmodStrategy (algo = "CEM", nbTryInInit = 50, nbTry=25)] (Lebret et al., 2015). The estimation of the mixture parameters can be carried out with a maximum likelihood using the EM algorithm (Expectation Maximization), the SEM (Stochastic EM), or by maximum likelihood classification using the CEM algorithm (Clustering EM). In this example, we use the EM algorithm. With the general family command, it is possible to give more flexibility to the model that best fits the data by allowing the volumes, shapes, and orientations of the groups to vary (Lebret et al., 2015).

Because our groups had different proportions (each group had a different sample size), the FALSE command was established as Z12 in the data frame. As an output of this estimation step, the program provides a partition and other parameters, including the proportions of the mixed model in each group, their averages, variances and likelihood, and the associated source of each group obtained for this example. Table 7 shows the output of the algorithm; in this table, it can be observed that although the analysis was carried out without labeling the observations, they were correctly assigned to their corresponding source group, except for a single observation of the Otumba (Ixtepec-Mailpais) subsource that was assigned to Otumba (Soltepec).

In this case, the model that best fitted the data turned out to be "Gaussian_pk_Lk_C" with the following cluster properties: Volume = Free, Shape = Equal, and Orientation = Equal. It is also possible to analyze the clustering results graphically. It is important to note that the graphics produced by the "ClusVis" package may vary in the output. The authors state that, for some specific reproducibility purposes, the Rmixmod package allows the random seed to be exactly controlled by providing the optional seed argument ("*set.seed: number*"). However, despite having performed multiple tests with different seeds, the resulting graphics tend to vary. Figure 14 displays the bivariate spherical Gaussian visualization associated with the confidence areas; the size of the gray areas around the centers reflects the size of the components. The accuracy of this representation is given by the difference between entropies and the percentage of inertia of the axes.

Table 7 Assignation of the mixing proportions with *Rmixmod* (z-partition of the obsidian sources).

Cluster	Proportion	Group assigned	n	Source
1	0.1269	6,6,6,6,6,6,6,6	8	Ahuisculco
2	0.0746	1,1,1,1,1,1,1,1,1,1,1,1,1,1,1,1,1	17	El Chayal
3	0.0522	4,4,4,4,4,4,4,4,4,4,4,4,4,4	14	San Martin Jilotepeque
4	0.1045	12,12,12,12,12,12,12,12,12,12,12, 12,12,12,12,12	16	Ixtepeque
5	0.0522	8,8,8,8,8,8,8,8,8,8,8	11	Otumba (Soltepec)
6	0.0597	7,7,7,7,7,7,7,7,7,7,7,7,7,7,7,7,7,7,7,7,7,7	22	Otumba (Ixtepec-Malpais)
7	0.1642	5,5,5,5,5,5,5	7	Oyameles
8	0.0821	11,11,11,11,11,11,11	7	Paredon
9	0.0746	10,10,10,10,10	5	Tulancingo-El Pizarrin
10	0.0373	2,2,2,2,2,2,2,2,2,2	10	Sierra de Pachuca
11	0.0522	9,9,9,9,9,9,9,9,9,9	10	Zacualtipan
12	0.1194	3,3,3,3,3,3,3	7	Zinapécuaro

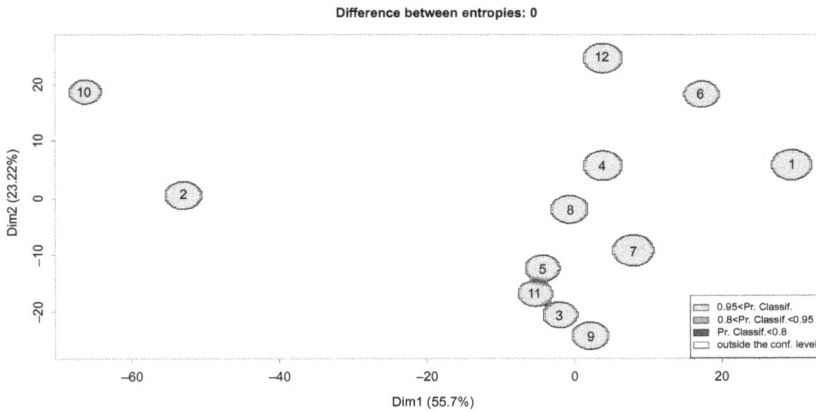

Figure 14 Component interpretation graph of the obsidian sources. Color version available at www.cambridge.org/argote_machine-learning

In the graph, it can be observed that the mapping of f is accurate because the difference between entropies is zero: $\delta_E(f, \hat{g}) = 0$. The first dimension provided by the LDA mapping was the most discriminative, with 55.7 percent of the discriminant power; the sum of the inertia of the first two axes was $55.7 + 23.22 = 78.92$ percent of the discriminant power; thus, most of the discriminant information was present on this two-dimensional mapping. Components 12, 7, 4, and 2 contain most of the observations. The components that show the greatest difference in their chemical composition are the samples from Components 1 and 6 (El Chayal and Ahuisculco, respectively) and Components 2 and 10 (Pachuca and Pizarrin) in the other extreme. Components 5 and 11 (Oyameles and El Paredon) are the ones that are closest to each other (regarding their mean vectors) and slightly join without meaning that the observations are mixed, as seen in the partition results of Table 7. An important observation about these results is that there are no overlaps between any of the sources used in the classification, hence fulfilling all the conditions of a good classification.

3.3 Exercise 2: Obsidian Sources in Guatemala

Carr (2015) performed a study to identify obsidian sources and subsources in the Guatemala Valley and the surrounding region. In his project, he analyzed a total of 215 samples from El Chayal, San Martin Jilotepeque, and Ixtepeque geological deposits with pXRF spectrometry. Of these samples, $n = 159$ were collected from 36 different localities in El Chayal, $n = 34$ were gathered from eight sampling localities in San Martin Jilotepeque, and $n = 22$ were collected from four sampling locations in Ixtepeque. The data matrix for this example can be found in the supplementary

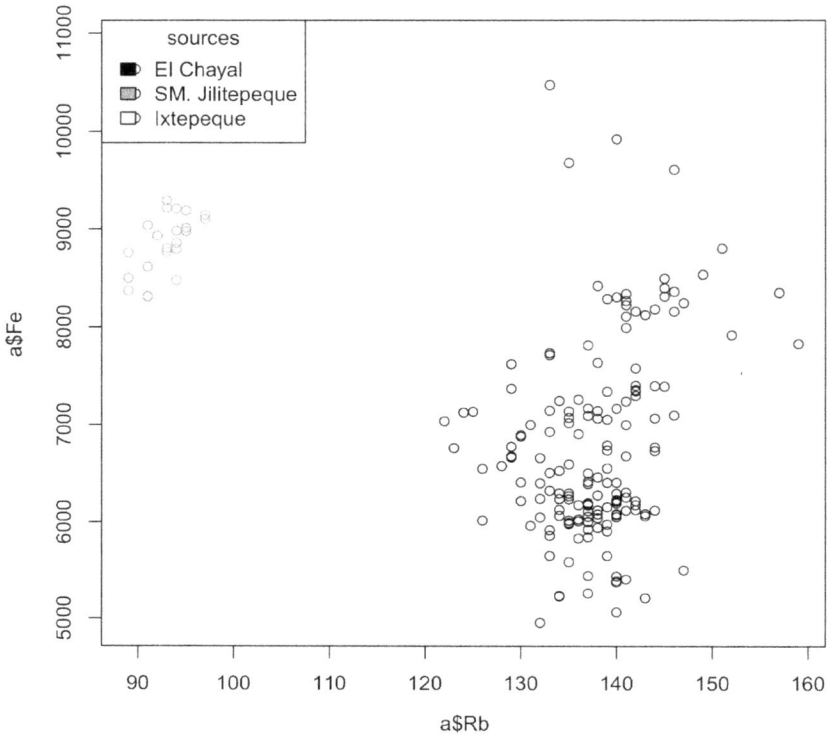

Figure 15 Bivariate plot using Rb (ppm) and Fe (ppm) concentrations of El Chayal, San Martin Jilotepeque, and Ixtepeque obsidian source systems (graphic reproduced with data from Carr, 2015). Color version available at www.cambridge.org/argote_machine-learning

file "Obs_maya.csv". To discriminate between these three main obsidian sources, Carr (2015) used bivariate graphs, plotting the concentrations of Rb *versus* Fe (see the graphic reproduced from his data in Figure 15). In the figure, it is possible to discriminate three different groups, but it is difficult to distinguish between different subsources. In addition, there is a great dispersion of the points from the El Chayal and Ixtepeque sources.

Carr (2015) also intended to examine the chemical variability of the samples to discriminate subsources within each of the main sources. For this purpose, the author analyzed the samples from each of the regions separately. For the El Chayal region, he used Rb *versus* Zr components in a bivariate display that resulted in five different subsources. Using this procedure, the author determined the existence of two distinct geochemical groups for the San Martin Jilotepeque obsidian source and two for Ixtepeque. To support the bivariate classification, Carr calculated the

Mahalanobis distances (MD) to obtain the group membership probabilities of the observations, making use of the following set of variables: Sr/Zr, Rb/Zr, Y/Zr, Fe/Mn, Mn, Fe, Zn, Ga, Rb, Sr, Y, Zr, Nb, and Th. It should be noted that this was not possible for El Chayal Subgroups 4 and 5 because their sample size was too small, preventing the calculation of their probabilities.

From his study, several observations can be made. First, the data were processed without any transformation. Therefore, it is advisable to open the data to remove the constant sum constraint. Second, the author managed to establish a total of nine obsidian subsources using bivariate graphs for each of the regions separately. Third, the calculation of probabilities to determine group memberships using the MD was not possible in all cases due to the restrictions imposed by the sample size. Furthermore, if Carr (2015) data were processed according to one of the commonly established methodologies, that is, transforming the data to log_{10} and applying a PCA, the explained variance of the first two components would have been only 56.53 percent, and six PCs would have been needed to explain 95.58 percent of the variance.

By plotting all the data of the nine subsources together using a PCA (Figure 16), including the information about the origin of the samples, it can be observed that the overlap between the different groups is unavoidable. In this way, it can be concluded that this methodology is not able to differentiate the chemical characteristics of the samples. Therefore, the discrimination of sources using concentrations is a procedure that requires nonconventional methods. According to this, we can assume that many of the published classifications that follow preestablished methods have incurred serious classification errors.

Employing the same data contained in Carr (2015), it was applied the methodology described in the associated section in *Statistical Processing of Quantitative Data of Archaeological Materials* www.cambridge.org/Argote. The entire procedure is described step by step in the supplementary video "*Video 4*". First, as part of the preprocessing, the data were transformed to the isometric log-ratio and then robustly standardized. Afterward, and as an essential step, the diagnosis of the data was made with the robust MCD estimator (see the scripts for this part of the procedure in the previous exercise). Figure 17 shows the distance-distance plot in which eleven outliers were detected (92, 115, 142, 150, 166, 189, 190, 191, 192, 193, 204), displayed in the right half of the graph. In this case, it was decided not to exclude them from the analysis because part of these observations (190 to 193) corresponded to the samples identified by Carr (2015) as belonging to the Jilotepeque 2 subsource (FP01, SAI01, SAI03, SAI04).

PROCESSING

COMPOSITIONAL

DATA

Exercise 2: Obsidian

Sources in

Guatemala

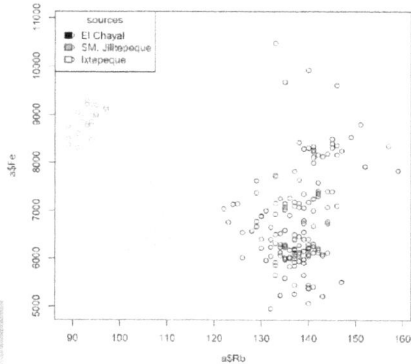

Video 4 Step-by-step video on how to compositional data of the obsidian samples used in Video 4. Video files available at www.cambridge.org/argote_machine-learning

Once the diagnosis was performed, unsupervised classification was carried out using the Gaussian mixture model with the *Rmixmod* and *clusVis* packages. The data matrix for this example consisted of $n = 215$ observations and $D = 10$ parts representing the Simplex. If the number of components is unknown and is to be estimated from the data, *Rmixmod* includes the parameter *nbCluster* to run a cluster analysis with a list of clusters (from 2 to n clusters). In this example, this parameter was set to "1" for the minimum number of components and "9" as the maximum number. Using the Gaussian mixture model, the Maximum likelihood inference was performed, and model selection was performed by the ICL criterion, detecting nine components. The output of the *Rmixmod* algorithm can be found in Table 8. The model that best fitted the data turned out to be Gaussian_pk_Lk_C, with clusters with the following properties: Volume = Free, Shape = Equal, and Orientation = Equal. The script used is the following:

```
rm(list=ls())
library(Rmixmod)
library(ClusVis)
data <- read.csv("C:\\Cobean\\Nueve_grupos\\robust_n-imp5.
csv",header=T)
str(data)
fammodel <- mixmodGaussianModel(family="general",equal.
proportions=FALSE)
Mod1<-mixmodCluster(data, 1:9, criterion = "ICL", strategy =
mixmodStrategy(algo = "CEM", nbTryInInit = 50, nbTry=25))
```

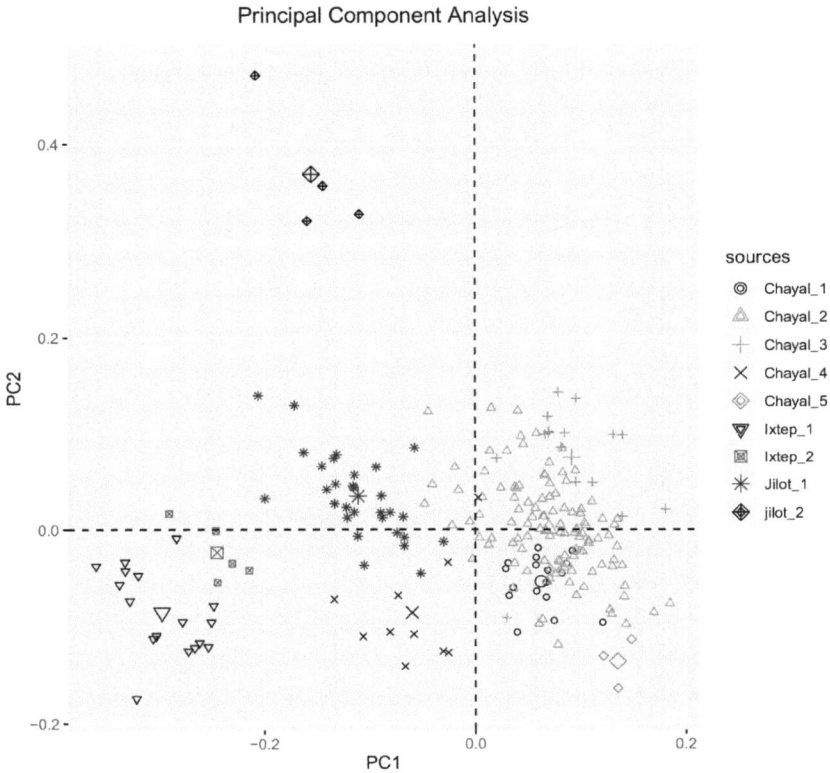

Figure 16 Plot of the two principal components of the samples from nine subsources in Guatemala valley. Color version available at www.cambridge.org/argote_machine-learning

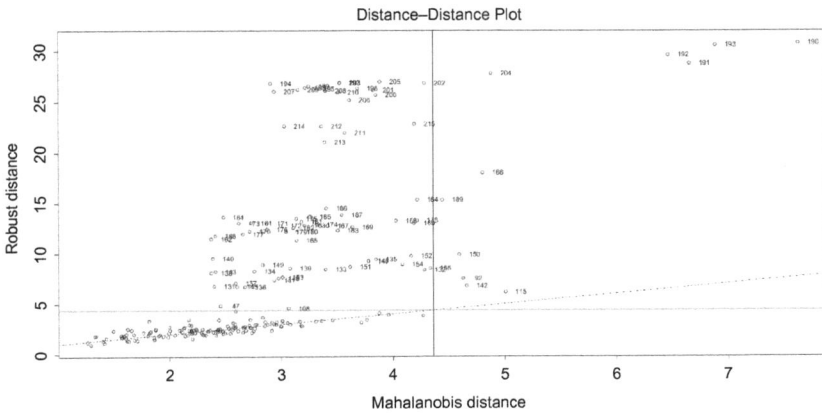

Figure 17 Distance–Distance plot of the samples of obsidian sources. Color version available at www.cambridge.org/argote_machine-learning

Table 8 Output of the *Rmixmod* algorithm.

```
*******************************************************

* Number of samples = 215
* Problem dimension = 9
*******************************************************

* Number of cluster = 9
* Model Type = Gaussian_pk_Lk_C
* Criterion = ICL(-3126.1685)
* Parameters = list by cluster
truncated output

z-partition
8 8 8 8 8 8 8 8 8 8 8 8 8 5 5 5 5 5 5 5 5 5 5 5 5 5 5 5 9 9 5 5 5 5 5 5 5 5 9 9 9 9 9 9 9 9 9 5 5
5 5 5 5 9 5 5 5 5 5 5 9 5 5 5 5 9 5 5 9 5 9 9 9 9 9 9 9 9 5 5 5 5 2 2 2 2 4 2 2 2 2 2 2 4
2 5 2 4 4 4 4 4 4 4 5 9 9 9 1 1 1 3 1 3 1 1 1 1 1 1 1 1 3 1 3 1 3 7 7 7 6 6 6 6 6 6 6 6 6 6
6 6
ICL(-3117.1847)
Log-likelihood = 1963.1254
```

```
summary(Mod1)
Mod1["partition"]
resvisu <- clusvisMixmod(Mod1)
plotDensityClusVisu(resvisu, add.obs = F, positionlegend =
"topleft")
```

The partition of the sample space of Figure 18 presents the Gaussian-like component overlap on the most discriminative map, where the difference between entropies is almost zero $[\delta_E(f, \hat{g}) = -0.01]$. The sum of the inertia of the first two axes is $61.62 + 18.93 = 80.55$ percent of the discriminant power. The groups with more observations are numbers 8, 3, 1, and 6. The most isolated groups are 1, 6, and 7; their position on opposite sides suggests that their chemical signatures are entirely different with respect to the rest of the subsources, with Component 7 differing the most. This component corresponds to Jilotepeque 2.

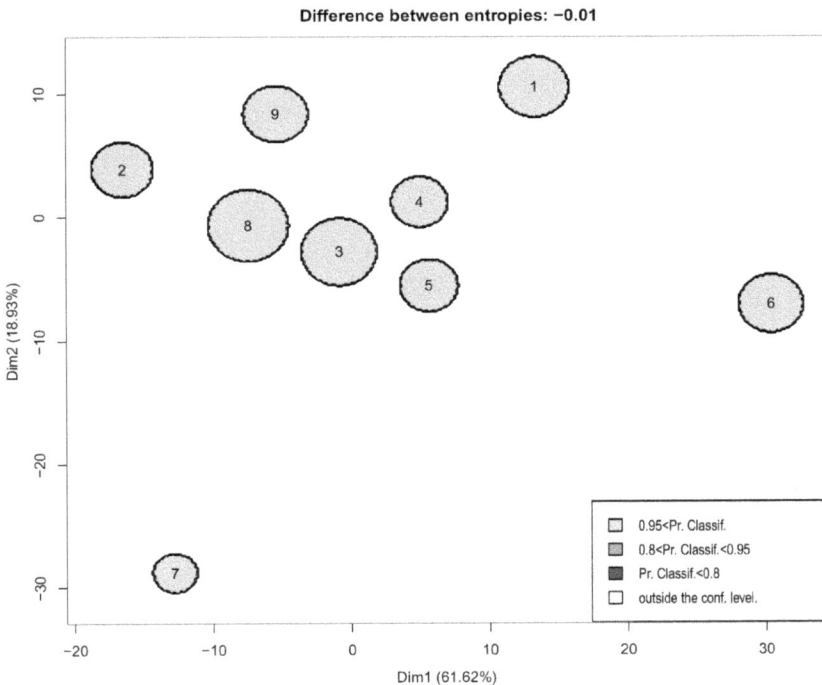

Figure 18 Component interpretation graph of the Mayan obsidian sources. The component numbers indicate the following subsources: [1] Jilotepeque 1, [2] El Chayal 3, [3] El Chayal 2*, [4] El Chayal 4, [5] Ixtepeque 2, [6] Ixtepeque 1, [7] Jilotepeque 2, [8] El Chayal 2, and [9] El Chayal 1. Color version available at www.cambridge.org/argote_machine-learning

The first thing that we can observe in these results is that the algorithm estimates the optimal number of groups as $K = 9$ components, which would correspond to the 9 subsources registered for the region. Therefore, it can be concluded that the chemical variability of each subsource is sufficiently different from the others, allowing recognition of its origin with adequate accuracy. However, the assignments of the samples to the components are different than the assignment made by Carr, as seen in the assignment of the units to each one of the components or subsources displayed in Table 9. This is because the mixture model is more robust for modeling group analysis.

Table 9 shows the z-partition of the Gaussian mixture model. In the first column are the nine subsources classified by Carr (2015). In the second column, the sample size of each subsource estimated by the author is tabulated. In the third column, the z-partition of nine components of the sample can be observed. The number of observations assigned to each component is recorded in the fourth column, and the proportions of each component are recorded in the fifth column. In the last column are the differences between the algorithm assignation and the classification provided by Carr.

In this example, the first component coincides with Carr's classification; that is, the same 17 sample units were assigned to the El Chayal 1 subsource. For the El Chayal 2 subsource, Carr classified 113 samples to this subsource, while the mixture model assigns only 78 observations to this component and 40

Table 9 Assignation of the mixture proportions with *Rmixmod* (z-partition of Guatemala obsidian sources and subsources).

Carr (2015) Subsource	Carr (2015) sample size (n)	Rmixmod component group ID	Rmixmod sample assignation	Proportion	z-partition
El Chayal 1	17	8	17	0.0791	El Chayal 1
El Chayal 2	113	5	78	0.3628	El Chayal 2
		9	40	0.1860	El Chayal 2*
El Chayal 3	16	2	13	0.0605	El Chayal 3
El Chayal 4	10	4	11	0.0512	El Chayal 4
El Chayal 5	3				
Jilotepeque 1	30	1	26	0.1209	Jilotepeque 1
		3	4	0.0186	Jilotepeque 1*
Jilotepeque 2	4	7	4	0.0186	Jilotepeque 2
Ixtepeque 1	17	6	22	0.1023	Ixtepeque 1
Ixtepeque 2	5				
Total =	**215**		**215**	**1.0000**	

Log-likelihood = 1963.1254 ICL(-3117.1847)

observations to component 9, now marked as El Chayal 2*. In other words, the original El Chayal 2 subsource was divided into two subsources by the mixture model, separating 35 samples from the original El Chayal 2 and transferring two samples from El Chayal 3 and three from El Chayal 5 to the new El Chayal 2* subgroup. Carr classified 16 observations in the El Chayal 3 group and 10 samples in the El Chayal 4 group, but the mixing model classification assigned only 13 samples to the El Chayal 3 group and 11 to the El Chayal 4 group. Due to the new assignations, the original El Chayal 5 subsource in Carr's classification was eliminated.

For Jilotepeque, Carr identified two subsources: Jilotepeque 1 and Jilotepeque 2, with 30 and 4 observations, respectively. The mixture model also identified two subsources but with slight differences in sample size for the case of Jilotepeque 1, with 26 observations in one component and 4 observations in another subsource marked as Jilotepeque 1*. The case of Jilotepeque 2 still consisted of four samples. For the Ixtepeque subsources, Carr identified two subsources: Ixtepeque 1 with 17 observations and Ixtepeque 2 with five observations. Conversely, the mixture model did not find significant differences to divide this source into two subsources, so the 22 original observations were assigned to a single component (Ixtepeque 1).

The Mahalanobis distance (MD) has been widely used as a classification technique in archaeometry to estimate relative probabilities of group membership. However, as discussed in the introduction of *Statistical Processing of Quantitative Data of Archaeological Materials* www.cambridge.org/Argote, this statistic presents several drawbacks that can cause serious problems in the calculation of group memberships. For example, Glascock et al. (1998) analyzed the provenance of obsidian samples from Central Mexico using the first three principal components, which explained approximately 92 percent of the variance. The author used the MD to calculate relative probabilities for the assignment of samples to the groups and obtained erroneous results for the sources of Santa Elena (Hidalgo) and the subsources of Pachuca. This exhibits that using a technique such as MD to make assignments can force some observations to belong to one of the groups. Furthermore, it was shown that multimodal distance distributions are preferable (Thrun, 2021b), which is a property that MD rarely possesses.

In contrast, model-based clustering produces an association weight based on a formal probability, called the posterior probability of each observation belonging to each group. The partition is derived from a maximum likelihood estimator using the MAP procedure (see the section on model-based clustering in *Statistical Processing of Quantitative Data of Archaeological Materials* www.cambridge.org/Argote). This procedure is carried out through the EM algorithm or one of its variants. Parameter estimation using the EM algorithm calculates the weights for each observation, given the parameter values of the mixture components and the overall mixture weights (Kessler, 2019). These weights are used as measures of the strength of the association between each observation and the groups in the model in such a way that, in the model-based method, the observations in the same cluster are generated from the same probability distribution of the cluster (Grün, 2019).

To exemplify this procedure, the groups established by Carr (2015) were contrasted according to his scatter diagrams and the probabilities obtained from the calculation of the MD made by the same author and the probabilities calculated using the mixture model. The reader can directly contrast the results of the calculation of the probabilities of membership obtained with both methods by directly consulting the work of Carr (2015) and computing the model-based clustering probabilities with the following command:

```
(Mod1@bestResult@proba) # Calculation of probabilities with
the mixture model
```

By making this comparison, one can appreciate that the memberships calculated with the MD do not coincide with the groups designated by Carr for each of the localities in the region. It is important to note that, in some cases, the MD underestimates and, in others, overestimates the probabilities of belonging to a group. The MD sometimes assigns the observations to the groups even when their probability of membership to a group is well below 50 percent. This fact has to do with the clear violation of the assumption of normality established by the MD method. On the other hand, the probabilities obtained by the mixture model are usually above 90 percent, except for three cases.

If the number of components is small and the sample size is large, the *ClusVis* program allows a pseudoscatter plot of the observation memberships to be obtained. In this plot, each observation is projected as colored dots representing the partition membership z; the information about the uncertainty of the classification is given by the curves of the *iso*-probability of classification, and information about the visualization accuracy is given by the difference between entropies and the percentages of inertia (Biernacki et al., 2021). In this case, it was considered to use only the five subsources belonging to El Chayal, forming a data matrix with $n = 159$ and $D = 10$; the processing was done the same way as for the previous case.

In Figure 19, we can see that the difference between entropies is almost zero $[\delta_E(f, \hat{g}) = -0.01]$. The sum of the inertia of the first two axes is $61.94 + 26.98 = 88.92$ percent of the discriminant power. Three probability levels of classification were obtained for El Chayal samples (0.50, 0.80, and 0.95); the observations are represented with the label of the component maximizing the posterior

Difference between entropies: −0.01

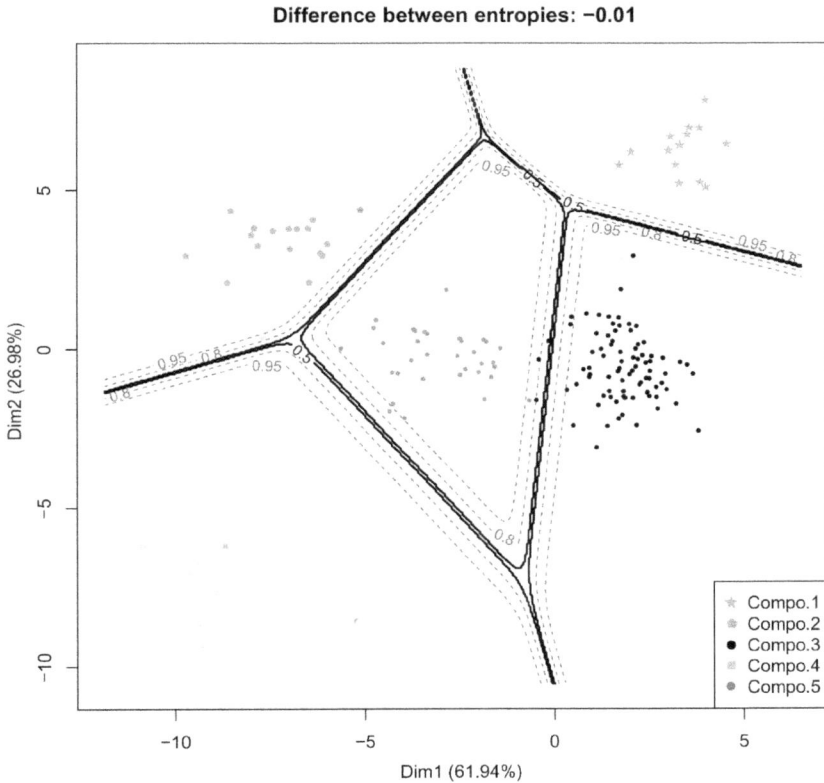

Figure 19 Bivariate scatter plot of the observation memberships of samples from El Chayal, Guatemala. Component 1 corresponds to El Chayal 3, component 2 to El Chayal 1, component 3 to El Chayal 2, component 4 to El Chayal 4, and component 5 to El Chayal 2 *. Color version available at www.cambridge.org/argote_machine-learning

probability of classification. Only four of the 159 observations were misclassified into component 4; in the rest of the components, there are no misclassified observations, and the membership of the observations is above 0.80 percent and 0.95 percent. To obtain the pseudoscatter plot of the observation memberships, the following command is used:

```
plotDensityClusVisu(resvisu, add.obs = T, positionle-
gend = " topleft")
```

At times it is important to have an idea of the chemical variation of each of the existing subsources and, for this, it is not enough to compare the vectors of means or their standard deviations. To obtain a better notion of the variability between subsources, one can turn to the definition of Binary Sequential Partition (SBP) of parts that defines an orthonormal basis in the Simplex (see the section on Compositional and Completely Compositional Data in *Statistical Processing of Quantitative Data of Archaeological Materials* www.cambridge.org/Argote). The partition criterion must be based on a wide knowledge of the association and affinity between parts (van den Boogaart and Tolosana-Delgado, 2013) but can also be obtained through the variation matrix of the log-ratios or the implementation of compositional biplots obtained from the variance and covariance matrices using the *clr* transformation (Pawlowsky-Glahn and Egozcue, 2011).

Another option is to group the parts by means of a hierarchical cluster analysis, such as the Ward method, and use the variation matrix to calculate the distances between the parts. With this method, a signed matrix is built, containing the bases from which an *ilr* definition matrix is obtained. The employment of the SBP allows the bases to be obtained to generate a CoDa-dendogram (Parent et al., 2012). To calculate the SBP and the CoDa-dendogram, it is necessary to include in the data matrix a column that identifies the group to which each of the samples belongs. As an example, the matrix in Table 10 illustrates the ordering of the groups.

To obtain the SBP, use this script:

```
rm(list=ls())
dat1 <- read.csv(X,header=T) #uploads the data file X arranged by
groups
str(dat1)#view the file structure
library(compositions)#uploads the package
x = acomp(dat1[,-c(1:2)]) #applies the closure operator only to
numeric data
x#to see the closed data
gr = dat1[,2] #assigns the variable class to gr= to identify the
obsidian groups (useful afterwards)
gr#to observe if the variable was correctly assigned
#Use an ilr basis coming from a clustering of parts
dd = dist(t(clr(x))) #computes the Euclidian distances of the
variation matrix
hc1 = hclust(dd,method="ward.D2") #builds the dendrogram with
the Ward method
plot(hc1)#dendogram
mergetree=hc1$merge#basis to use, described as a merging tree
color=c("green3","darkviolet","red", "blue", "orange")
CoDaDendrogram(X=acomp(x),mergetree=mergetree,col="black",
range=c(-6,6),type="l")
xsplit = split(x,gr)
for(i in 1:5){
```

Table 10 Example of the matrix to obtain the SPB bases and the CoDa-dendogram.

ID	Class	Mn	Fe	Zn	Ga	Th	Rb	Sr	Y	Zr	Nb
CH01	group2	651	5844	30	17	9	137	133	17	101	9
CH02	group2	712	6075	32	17	10	140	139	20	105	9
CH03	group2	687	5943	33	17	10	138	138	20	106	10
CH04	group2	662	6009	34	17	9	136	135	19	105	9
CH05	group2	680	6029	37	18	10	136	139	21	107	9
CH06	group2	656	6084	34	17	12	140	140	19	106	9
CH07	group2	679	6115	33	17	10	137	138	18	106	10
CH08	group2	667	5994	30	17	9	135	137	19	108	10
CH09	group2	656	6042	33	17	10	138	134	19	105	9
CH10	group2	617	6155	34	17	10	139	140	20	103	10
LM13	group2	680	6399	31	17	8	132	132	18	105	9
LJ01	group5	703	7829	36	18	11	159	165	22	120	12
LJ02	group5	668	7920	35	18	12	152	161	20	116	10
LJ03	group5	651	8352	43	18	12	157	172	23	115	9
AC13	group5	516	6675	29	17	10	129	125	18	103	7
KM01	group5	597	6999	37	18	11	141	149	19	107	9
KM02	group5	712	7243	31	17	10	141	153	21	111	9
KM03	group5	669	8248	32	17	10	147	154	20	109	10

```
CoDaDendrogram(X=acomp(xsplit[[i]]),border=color[i],
type="box",box.pos=i-2.5,box.space=1.5,add=TRUE)
CoDaDendrogram(X=acomp(xsplit[[i]]),col=color[i],
type="line",add=TRUE)
}
```

The CoDa-dendogram of the El Chayal dataset, following the SBP, is represented in Figure 20. In the electronic version of this Element, the colors assigned to the subsources can be appreciated, which were the following: El Chayal 1 = red, El Chayal 2* = green, El Chayal 2 = purple, El Chayal 3 = blue and El Chayal 4 = orange. The length of the vertical bars represents the variability of each *ilr* coordinate. In this case, we have nine balances with $D = 10$ parts (components). As seen, the third balance has the largest variance and involves the Fe (iron) part; this implies that this balance explains a great portion of the total variance. The location of the mean of an *ilr*-coordinate is determined by the intersection of the vertical segment with the horizontal segment (variance); when this intersection is not centered, it indicates a major or minor influence of one of the groups of the parts.

For this example, we can see in the first balance (*b1*) that the median of the five groups does not coincide. In the first balance, the box plot shows a certain asymmetry and a greater dispersion than the second balance. This is due to the low variability given by Fe, Mn, Zr, Rb, and Sr concentrations in the second group of parts. In balance 1 (*b1*), El Chayal 4 (orange) deviates slightly from the other groups and shows a greater contribution in the Fe–Mn–Zr–Rb–Sr parts. Conversely, El Chayal 2* (green) shows smaller amounts in the Th–Nb–Zn–Ga–Y parts. The other groups contain slight variations between the two major groups of the parts. In balance 2 (*b2*), El Chayal 1 (red) has a lower variance, and as it is loaded to the left, it would have smaller quantities of Th and Nb parts and a greater contribution of Zn–Ga–Y parts.

In balance 3 (*b3*), El Chayal 2 (purple) shows a lower content of Fe and minimal variations in Mn-Zr-Rb–Sr with respect to the other groups. Balance 7 (*b7*) also displays a similar aspect, but El Chayal 1 shows a slight increase in the Zr, Rb, and Sr parts, followed by El Chayal 2* and El Chayal 4. The effect is null in the rest of the balances because there is good symmetry in the parts. El Chayal 4 registers a greater variance in balances *b4* and *b5*, as does El Chayal 3, so these two subsources would present a greater dispersion in this group of parts (Th–Nb–Zn–Ga–Y). The parts that play a greater role in the classification are those to the right of the CoDa-dendrogram (Fe–Mn–Zr–Rb–Sr).

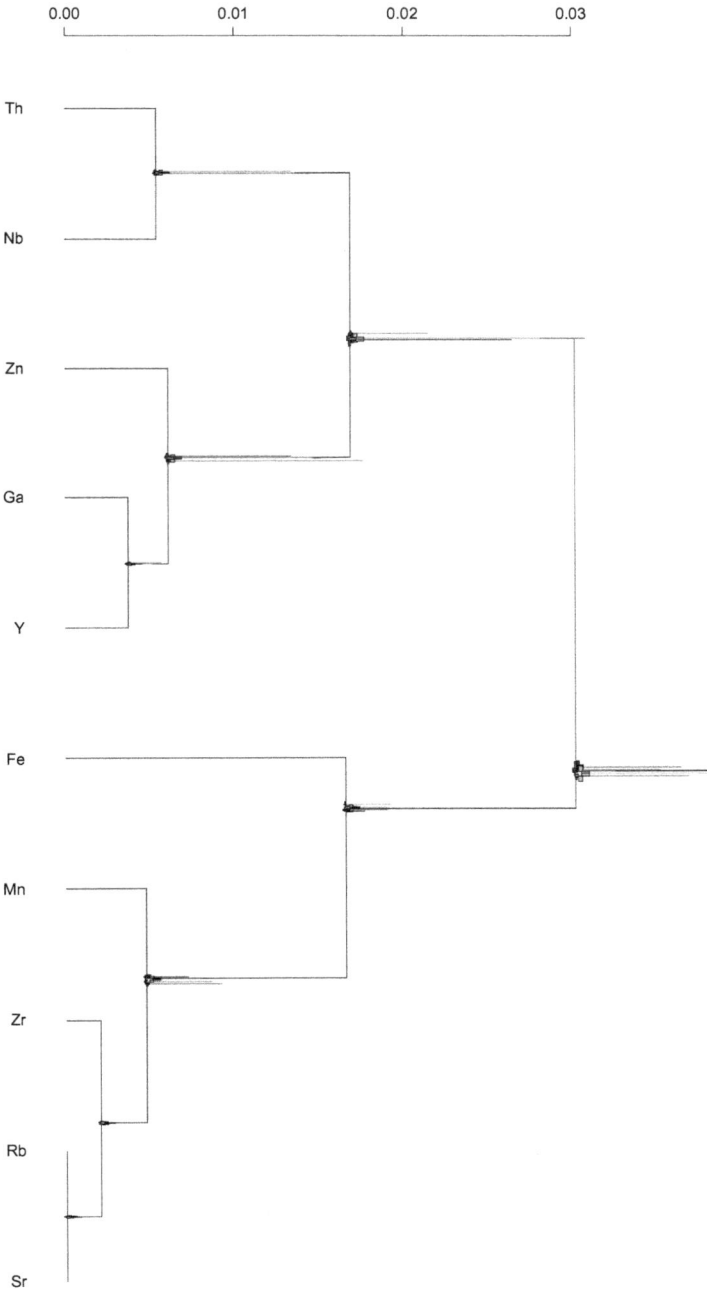

Figure 20 CoDa-dendrogram for the El Chayal dataset using SBP. The vertical bars correspond to the decomposition of the variance by balances. Color version available at www.cambridge.org/argote_machine-learning

In summary, it can be said that the range of chemical variability of each of the subsources is different and that the discrimination between them is because the material does not correspond to the same lava flow, which would explain the variation in the chemical elements that compose the samples. In general, one can speak of the existence of five different systems that could correspond to different eruptions or volcanic events.

4 Processing a Combination of Spectral and Compositional Data

4.1 Application and Case Studies

Let us remember that chemometry is directly related to all those methods that transform relatively complex analytical signals and data to provide the maximum amount of relevant chemical information; it is strongly connected to multivariate quantitative analysis and pattern recognition. From the point of view of chemometry, spectral data can be represented in a matrix form (Figure 21) for further classification with multivariate methods. This matrix (i.e., spectral data) combined with a compositional data vector can be used for the classification of archaeological samples combining spectral preprocessing techniques (such as that described in Section 2 of this Element), variable selection methods, and projection-based clustering analysis. In this section, the proposed methods will be applied to data measured from obsidian samples and to a hypothetical "human-in-the-loop" cluster analysis

4.2 Exercise 1: Mesoamerican Obsidian Deposits

In this example, geological samples from eight different obsidian sources (enlisted in Table 11) were analyzed with a pXRF spectrometer, employing a TRACER III-SD XRF portable analyzer manufactured by Bruker Corporation, with an Rh tube at an angle of 52°, a drift silicon detector, a 7.5 μm Be detector window and a factory filter composed of 6 μm Cu, 1 μm Tl, and 12 μm Al. The instrument was set with a

variables

$$A = \begin{vmatrix} a_{11} & a_{12} & a_{13} & \dots & a_{1n} \\ a_{21} & a_{22} & a_{23} & \dots & a_{2n} \\ a_{31} & a_{32} & a_{33} & \dots & a_{3n} \\ \vdots & \vdots & \vdots & \ddots & \vdots \\ a_{m1} & a_{m2} & a_{m3} & \dots & a_{mn} \end{vmatrix} \quad \text{observations}$$

Figure 21 Matrix representation of the spectra. Color version available at www.cambridge.org/argote_machine-learning

Table 11 Number of samples collected from eight Mesoamerican obsidian sources.

Source name	Geographic region	*n*
Ahuisculco	Jalisco	9
El Chayal	Guatemala	34
San Martin Jilotepeque	Guatemala	17
Ixtepeque	Guatemala	17
La Esperanza	Honduras	16
Otumba (Soltepec)	Edo. de México	10
Otumba (Ixtepec-Malpais)	Edo. de México	23
Sierra de Pachuca	Hidalgo	10
	Total =	136

voltage of 40 kV, a current of 30 μA, and a measurement time of 200 live seconds. The measured spectra were used to construct a matrix that consisted of $n = 136$ [samples] and $p = 2,048$ [channels], available in the supplementary material as file "Obsidian_chp4.csv". The spectral and compositional data from these sources are similar to the data used in the first exercise of Sections 2 and 3 of this Element. The procedure is described step by step in the supplementary video "*Video 5.*"

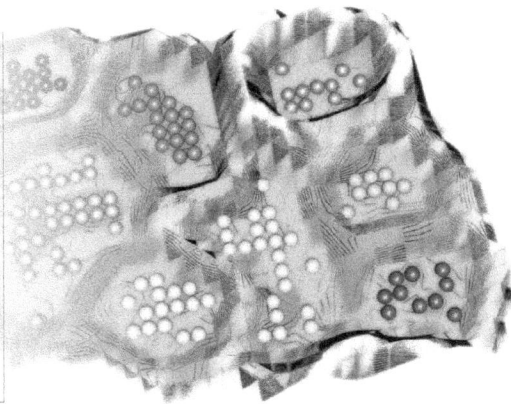

Video 5 Step-by-step video on how to process a combination of spectral and compositional data used in Video 5. Video files available at www.cambridge.org/argote_machine-learning

In this case, the spectra showed a high number of noisy signals that are more visible at higher keV values (Figure 22). Moreover, at the beginning and the end

Full spectrum

Selected region

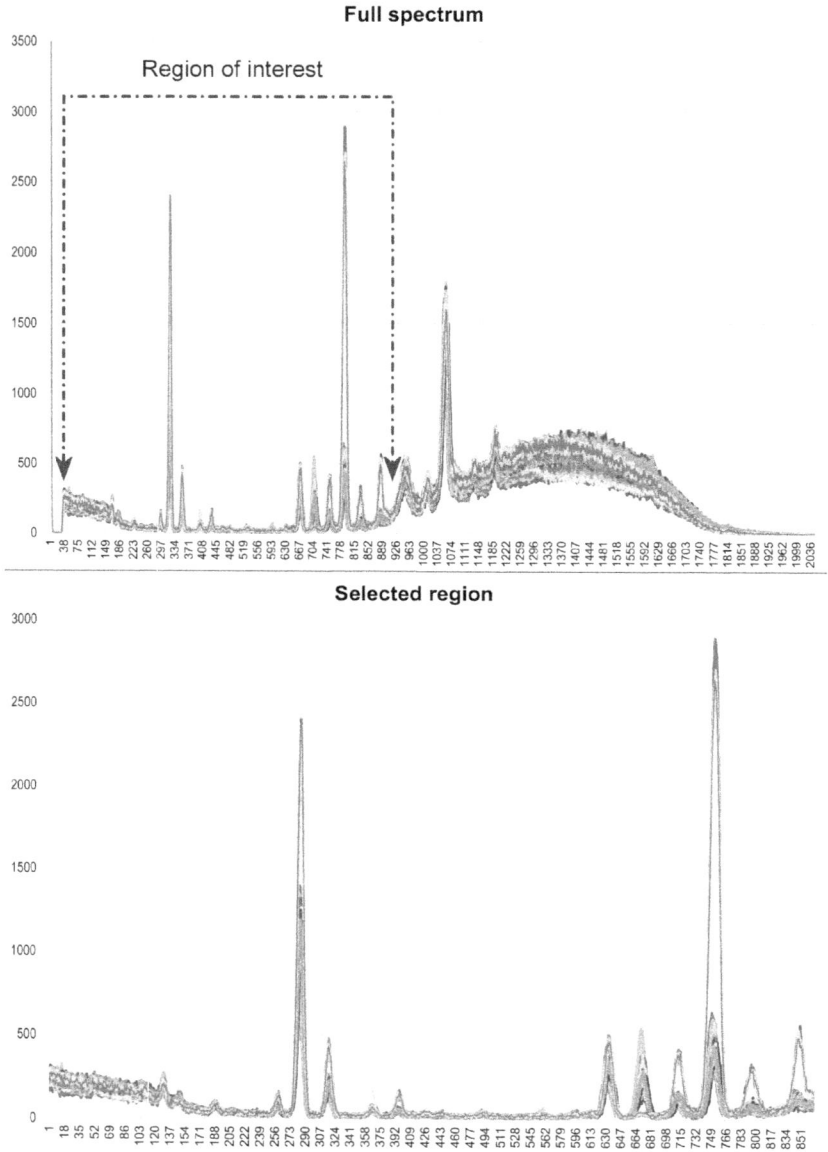

Figure 22 (Top) Full spectra of the obsidian samples. (Bottom) Selection of useful channels from the *p*XRF spectra. Color version available at www.cambridge.org/argote_machine-learning

of the spectra are areas that contain no relevant information, either because they contain values close to zero or correspond to undesirable effects such as the Compton peak or Raleigh scattering. Therefore, columns 1 to 38 and 903 to

2,048 were manually deleted from the original matrix, reducing the dimension of the new matrix to $n = 136$ obsidian samples by $p = 865$ channel counts (matrix available in the supplementary material as file 'Ixtq_2_38_865.csv').

The second step was to filter the spectra to maximize the quality of the information. The raw spectra were processed using the EMSC and SG filters separately; the two resulting matrices were saved in different files. Afterward, the EMSC-filtered data were processed using the Savitzky–Golay filter, and the SG-filtered matrix was treated with the EMSC algorithm to obtain two other files, one with a combination of EMSC + SG and one with SG + EMSC filters. It is important to note that these procedures eliminate five columns from both extremes of the matrix, leaving only 855 variables in our final matrix. The filtering system with the best performance is chosen by evaluating the parameter values that are calculated later in the procedure. The script below allows filtering with the EMSC algorithm and then filtering with the SG algorithm. Note that to perform the inverse action, first use the script for the SG filter and afterward the code for the EMSC. For individual filtering systems, that is, only EMSC or only SG, the scripts are used separately. Just remember to update the file names to call and run the proper one.

```
## Script to filter with the EMSC algorithm
rm(list = ls())
library(EMSC) #Package EMSC. Performs model-based background
correction and
# normalization of the spectra (Liland and Indahl, 2020)
dat <- read.csv("C: \\Ixtq_2_38_865.csv'", header=T) # To call
the spectral data file
str(dat) # to see the data structure
dat1 <- dat[,2:866] # To eliminate the first column related to the
sample identifier
str(dat1)
EMSC.basic <- EMSC(dat1)
EMSC.poly6 <- EMSC(dat1, degree = 6) #Filters the spectra with a
6th order
# polynomial
str(EMSC.poly6)
write.csv(EMSC.poly6$corrected, file="FEmsc.csv") # to save
the data file
# filtered with the EMSC. The user can choose other file names
##---------------------------
## Script to filter with the SG algorithm
library(prospectr) ## Miscellaneous Functions for Processing
and Sample Selection of
```

```
## Spectroscopic Data (Stevens et al., 2022)
data <- read.csv("FEmsc.csv", header=T) # Calls the file with
EMSC filtered data
# data <- read.csv("C: \\Ixtq_2_38_865.csv'", header=T) #Use
this instead for only SG
str(data)
# data1 <- dat[,2:866] #Add this when applying only SG
#str(data1) #Add this when applying only SG
sg <- savitzkyGolay(data, p = 3, w = 11, m = 0)
# sg <- savitzkyGolay(data1, p = 3, w = 11, m = 0) #Use this instead
for only SG
write.csv(sg, file=" FEmsc _SG.csv") # or 'SG.csv'. The user can
choose other file names
```

In the third step, the data were processed using the iPLS algorithm to select the interval(s) within the data that would provide the most significant variables. In the model, the photon counts in each of the spectrum channels measured with the pXRF instrument were used as the explanatory variables (X). To obtain the response variable (y), the luminescent data were converted to concentration values according to the Empirical Coefficients method (Rowe et al., 2012) using a variant of the Lukas-Tooth and Price (1961) equation. Once the chemical concentrations were obtained, the resulting matrix contained the following components: Mn, Fe, Zn, Ga, Th, Rb, Sr, Y, Zr, and Nb; this matrix is provided in the supplementary material of the electronic version of this volume as file "analitos.csv". The data were then moved to their native geometric space according to Aitchison's theory (Aitchison, 1986) employing the centered log-ratio (*clr*) transformation using the following script:

```
rm(list=ls())
data <- read.csv ("C:\\analitos.csv", header=T) #You can use
your own file
str(data) # displays the internal structure of the file, which
includes the format of each#column
dat2 <- data[2:11] # delete data identification column
str(dat2)
# transformation of the data to the clr log-ratio
library(compositions) ## van den Boogaart, Tolosana-Delgado
and Bren (2023)
xxat1 <- acomp(dat2) #the function "acomp" representing one
closed composition.
#With this command, the dataset is now closed.
xxat2 <- clr(xxat1) # Centered logratio transformation
str(xxat2)
```

```
write.csv(xxat2, file="Clr-transformation.csv") ##From this
resulting output file
#("Clr-transformation.csv"), extract Sr (strontium) vector
and save it as a separate.cvs
# file
```

From the resulting output file, the values obtained for the variable Sr (strontium) are copied to a separate .cvs file for its posterior use in the iPLS calibration model as the response variable ($y = $ Sr). It is important to note that the selection of the response variable will depend on the material you are working with. For example, pottery, Fe or Ca could be relevant; for some Roman glasses, Na, Ca, Sb, and Pb can differentiate the real origin of the pieces (López-García and Argote, 2023); for obsidian, Sr is a discriminatory element, so it was chosen for the analysis. For this exercise, the file "Stroncio.csv" is provided in the supplementary material. Table 12 shows an example of the difference between the raw values of Sr composition of some samples and its *clr* transformed values.

For iPLS regression analysis, the full spectrum (1–855) was divided into 10 equidistant subintervals, each containing approximately 85 variables. Then, a PLS calibration model was developed for each subinterval. The iPLS algorithm was applied to the four filtering systems (EMSC, SG, ESMC+SG, and SG +EMSC), and the one with the best performance was chosen according to the

Table 12 Example of the raw compositional data and the data after the *clr* transformation.

ID	Raw data Sr	Clr transformation Sr
1	129.36	0.465173738
2	132.46	0.507479359
3	131.64	0.504697572
4	132.17	0.474821295
5	127.27	0.439785067
6	133.33	0.488857386
7	137.82	0.502847437
134	4.61	−3.303411956
135	5.60	−3.137748288
136	7.23	−2.938981779

values observed in the RMSECV, RMSE, and R^2 parameters. The EMSC + SG combination turns out to be the best for filtering the data. To run the iPLS algorithm using the next script, the output was the following:

```
## Interval variable selection
rm(list=ls()) ## To remove all objects from memory
library(mdatools) # Kucheryavskiy (2020)
X <- read.csv("C:\\FEmsc_SG.csv", header=T) #Call the filtered
spectra data file
str(X)
# mean centering, in case you want to autoscale the spectrum
#X1 = prep.autoscale(X, center = T, scale = F)
# Call the concentration data file of Sr ("Stroncio.csv") or the y
variable you selected
y <- read.csv("C:\\Stroncio.csv", header=T)
str(y)
# ipls model
# for a model without mean centering, use X instead of X1
m = ipls(X, y, glob.ncomp = 4, int.num = 10)
# Model parameters
# glob.ncomp = maximum number of components for a global PLS
model
# int.num = number of intervals
summary(m)
plot(m)
plotRMSE(m)
show(m$int.selected)
show(m$var.selected)
par(mfrow = c(1, 2))
```

The output details information about the selected intervals, the number of variables at both ends, and the value of R^2:

```
Model with all intervals: RMSECV = 0.107, nLV = 4
Iteration 1/ 10 ... selected interval 9 (RMSECV = 0.125, nLV = 4)
Iteration 2/ 10 ... selected interval 8 (RMSECV = 0.105, nLV = 4)
Iteration 3/ 10 ... selected interval 2 (RMSECV = 0.105, nLV = 4)
Iteration 4/ 10 ... selected interval 5 (RMSECV = 0.105, nLV = 4)
Iteration 5/ 10 ... selected interval 3 (RMSECV = 0.105, nLV = 4)
Iteration 6/ 10 ... selected interval 1 (RMSECV = 0.105, nLV = 4)
Iteration 7/ 10 ... no improvements, stop.
```

In this case, the global model has an RMSECV = 0.107 with four components. Interval 9, with RMSECV = 0.125, gave the best performance for building local models with individual intervals. On the other hand, the second iteration selected interval 8 as the best local model with an RMSECV = 0.105, and according to the cross-validation, four components were optimal. In the subsequent iterations, more intervals were included; nevertheless, the RMSECV value did not vary at all, so it was no longer convenient to add more intervals. Intervals 8 and 9 present an R^2 similar to the global PLS model using the entire spectrum, suggesting a meaningless variation of the RMSE or R^2 values if any other intervals were included. In this way, it was possible to determine that intervals 8 and 9 contained the most informative variables of the spectrum; these intervals included the range of variables from columns 601 to 685 and from 686 to 770, respectively.

The iPLS variable selection results were as follows:

```
Validation: venetian blinds with 10 segments
Number of intervals: 10
Number of selected intervals: 6
RMSECV for global model: 0.107508 (4 LVs)
RMSECV for optimized model: 0.105035 (4 LVs)
Summary for selection procedure:
n start end selected nComp RMSE R2
1 0 1 855 FALSE 4 0.107 0.988 Global Model
2 9 686 770 TRUE 4 0.125 0.984
3 8 601 685 TRUE 4 0.105 0.988
4 2 87 172 TRUE 4 0.105 0.988
show(m$int.selected)
[1] 9 8 2 5 3 1
Method: forward
```

Figure 23 shows the performance of individual models and the selected interval or intervals, interpreted as follows. The average spectrum can be appreciated along the bars. Green or dark gray bars are the local intervals selected by the iPLS model; the height of each bar corresponds to the RMSECV value for the local model using variables from this interval as predictors (X). The number within each bar is the number of PLS components used in the local model. A dashed horizontal line indicates the RMSECV obtained by using all variables, and the number 4 at the end of that line is the number of latent variables (LV).

Finally, it is convenient to check the existence of outliers using the ROBPCA algorithm of the 'rrcov' package:

```
rm(list=ls())
library(rrcov) ## Scalable Robust Estimators with High
```

iPLS results

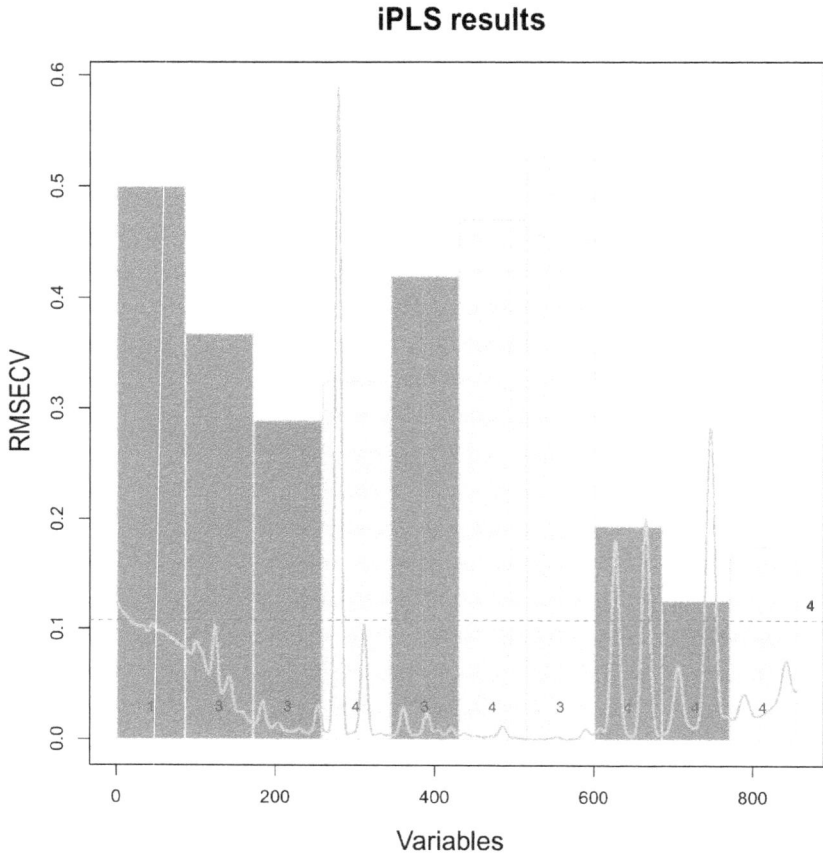

Figure 23 iPLS model for 10 intervals. Color version available at
www.cambridge.org/argote_machine-learning

```
Breakdown Point (Todorov, 2020)
dat <- read.csv("C:\\FEmsc_SG.csv", header=T)
str(dat)
rpca <- PcaGrid(~., data=dat)
rpca
plot(PcaHubert(dat, k=0), sub="data set: dat, k=4")
str(rpca)
rpca$flag
```

In this case, to establish the optimal number of components to retain, we set $k = 0$ such that $l_k / l_1 >= 10. E - 3$ and $\sum_{j=1}^{k} lj / \sum_{j=1}^{r} lj >= 0.8$. Refer to Hubert et al. (2005) and the 'rrcov' package (Todorov, 2020) in R for more information.

Robust PCA

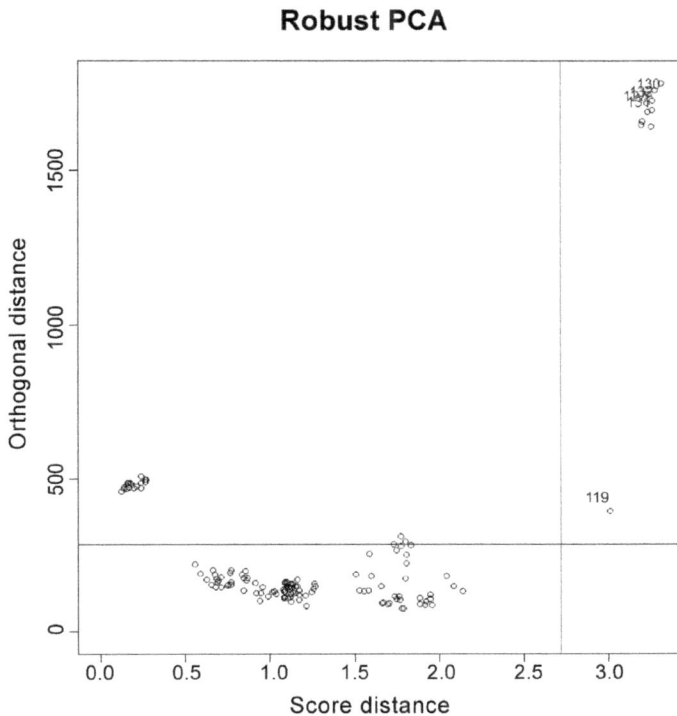

Figure 24 Robust diagnostic plot for the obsidian data with k = 4. Color version available at www.cambridge.org/argote_machine-learning

In the diagnosis of the data (Figure 24), eleven observations exceeded the cutoff value: ten forming a small group of bad leverage points at the top of the quadrant and the isolated case 119 at the far right of the graph. The bad leverage points correspond to the Pachuca samples; their separation from the rest of the sources is because these samples register significantly higher concentrations of Fe, Zr, and Zn and lower amounts of Sr compared to the other deposits. Therefore, they cannot be considered outliers. Conversely, the isolated sample (no. 119) belongs to Otumba (Ixtepec-Malpais); it is possible that it has some measurement error or contamination because it completely departs from the group of normal observations; thus, it is convenient to eliminate it from the analysis.

Now that the preprocessing has concluded, the next stage of the analysis is to perform the projection-based clustering. For this, we used a reduced matrix of $n = 135$ by $p = 170$ that included only the range from 601 to 770 channel counts (intervals 8 and 9 selected by the iPLS) and eliminated sample 119.

Although, in fact, the provenance of the sources is known, we considered this exercise as a nonsupervised classification, that is, where the data are unlabeled. The first thing we want to know is whether the data exhibit a clustering structure according to the information recorded in the variables. If this is true, then we will need to determine the optimal number of groups and the correct assignment of the observations using the projection-based clustering method (Thrun, 2018).

The first module of the script (see "Introduction into the Usage of Projection-based clustering" in *Statistical Processing of Quantitative Data of Archaeological Materials* www.cambridge.org/Argote) allows us to visually appreciate the existence of groups and determine the optimal number of these through a topographic map (Thrun et al., 2016). The colors presented by the topographic elements depend on their elevation and are based on the U-matrix principle (Ultsch, 2003; Ultsch and Siemon, 1990). The greater the spacing of the partitions in the high dimensional space, the higher the mountains on the topographic map. If two high-dimensional data points are in the same partition, both points end up in a blue lake or on a green meadow. Blue lakes indicate partitions with particularly high densities. Green grasslands represent homogeneous partitions. Conversely, if the two data points belong to different partitions, the landscape folds and a mountain range is created between the two points; depending on the real distance, the ranges can go from brown to white. Outliers land on volcanoes or mountaintops.

Here, we assume that the data was preprocessed appropriately and that the Euclidean distance is the best choice of similarity. Then the following code can be used:

```
rm(list = ls())
library(DatabionicSwarm)
datos <- read.csv("C:\\Two_intervals.csv", header=T)
str(datos)
datos=as.matrix(datos)

library(DatabionicSwarm)
InputDistances = as.matrix(dist(datos))
projection = Pswarm(InputDistances)
library(DatabionicSwarm)
library(GeneralizedUmatrix)
genUmatrixList=GeneratePswarmVisualization(
Data=datos,
projection$ProjectedPoints,
projection$LC)
```

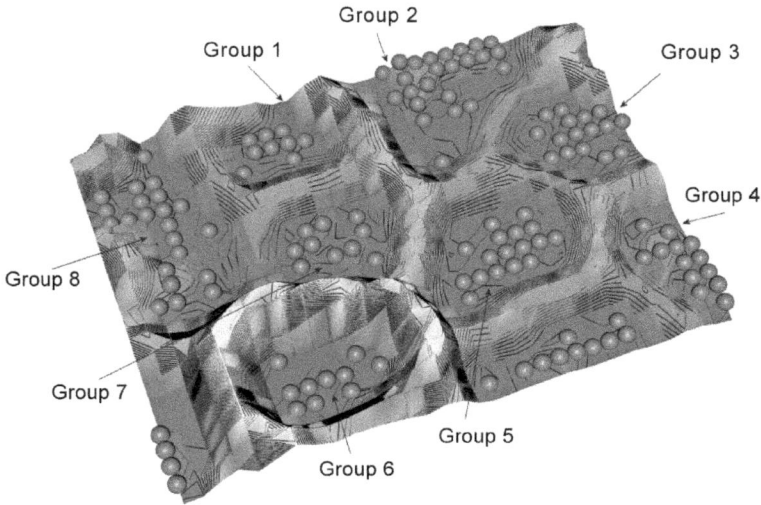

Figure 25 Topographic map of the DBS projection of the obsidian dataset with the Generalized U-matrix. Color version available at www.cambridge.org/argote_machine-learning

```
GeneralizedUmatrix::plotTopographicMap(
genUmatrixList$Umatrix,
genUmatrixList$Bestmatches,
NoLevels = 10)
```

Figure 25 displays the topographic map computed for the obsidian samples, where we can identify the existence of eight groups well separated by topographical barriers. The observations are clustered according to the position of the projected points. In this way, the points in each of the sections of the map correspond to observations that are similar between them and different from the rest of the groups according to their characteristics. In contrast to other multivariate methods, there is no overlap of the groups.

Once the existence of groups in the data is determined, the number of groups is specified in the script (with $k = 8$), and it is run again:

```
#Use previously loaded data
library(DatabionicSwarm) ## (Thrun, 2021a)
library(GeneralizedUmatrix) ## (Thrun et al., 2021a)
Cls = DBSclustering(
k = 8,
InputDistances,
genUmatrixList$Bestmatches,
```

```
genUmatrixList$LC,
PlotIt = FALSE
)
GeneralizedUmatrix::plotTopographicMap(

genUmatrixList$Umatrix,
genUmatrixList$Bestmatches,
Cls,
NoLevels = 10)
library(DatabionicSwarm)
library(ProjectionBasedClustering)
library(GeneralizedUmatrix)
   Imx = ProjectionBasedClustering::interactiveGeneralized
   UmatrixIsland(
   genUmatrixList$Umatrix,
   genUmatrixList$Bestmatches,
   Cls)
GeneralizedUmatrix::plotTopographicMap(
   genUmatrixList$Umatrix,
   genUmatrixList$Bestmatches,
   Cls = Cls,
   Imx = Imx)
   Cls # provides the labels of the instances
```

In this first part of the analysis, it is important to remember that the data continue to be processed as a nonsupervised classification, where there is no *a priori* information of the data. By setting $k = 8$, the algorithm labels the observations according to the distance and the density of each data point, assigning a different color to each group. In this way, a group can be distinguished from the others by their high-dimensional distances from the original dataset on a map by hypsometric tints defined by the generalized U matrix (see Figure 26). What is important in this map is that the groups are clearly visible with a more compact structure defined by the valleys. Similarly, it is possible to perceive the absence of outliers in the data.

In the last code, the "Cls" command allows you to see to which group was assigned each sample, as in Table 13.

In this example, all the samples were correctly assigned to their respective sources. As seen in Table 13, the first $n = 34$ samples were assigned to group 1 related to the El Chayal source, observations 35 to 50 ($n = 16$) were assigned to group 2 related to the La Esperanza source (Honduras), and so on. Therefore, it can be concluded that both the filtering system and the selection of the most informative intervals were appropriate procedures for partitioning the data into

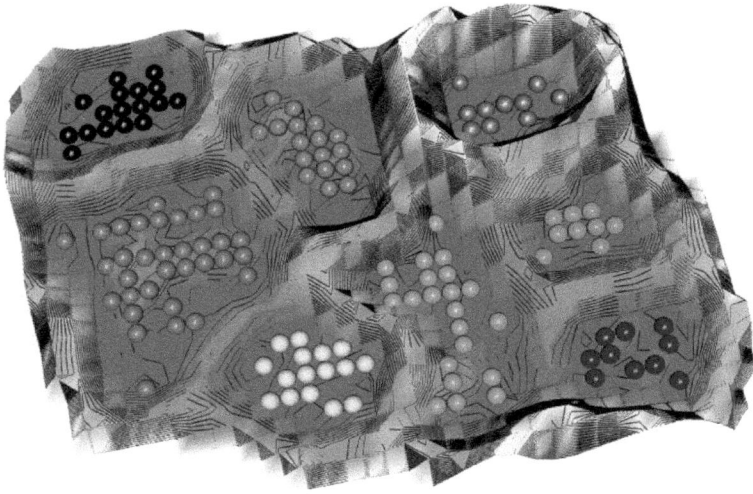

Figure 26 Topographic map of the DBS projection of the obsidian dataset with Generalized U-matrix using k = 8, interactively cropped. Color version available at www.cambridge.org/argote_machine-learning

natural or significant groups. The final step is validating the model. As mentioned before in this Element, it is important to confirm the obtained results in a quantitative and objective manner to establish if the model fits the data well or if only represents a spurious solution. One way is to calculate the percentage of the accuracy, a supervised index defined by the ratio of the number of true positives to the number of cases (see Section 4 in *Statistical Processing of Quantitative Data of Archaeological Materials* www.cambridge.org/Argote). Another approach is to validate a clustering with the help of domain experts (e.g., López-García et al., 2020; Thrun et al., 2021b; Thrun et al., 2022). A third option is to evaluate if clustering is useful for a specific application (Thrun, 2022).

For simplicity, we compute the cluster accuracy in the dataset. For this, it is necessary to add a column in the data matrix with the header "Cls", which refers to the class or group assigned by the algorithm to each of the observations (use the assignations of Table 13). Once this column has been added to the data matrix, the algorithm is rerun with the following script:

```
## with labeled data
rm(list=ls())
library(DatabionicSwarm) ## Thrun (2021a)
DataRaw <- read.csv("C:\\Two_intervals_cls.csv", header=T)
##Call the csv file with the #two intervals and add the Cls column.
The user can choose a personalized file name
```

Table 13 Assignation of samples to groups.

1	2	3	4	5	6	7	8	9	10	11	12	13	14	15	16	17	18
1	1	1	1	1	1	1	1	1	1	1	1	1	1	1	1	1	1
1	1	1	1	1	1	1	1	1	1	1	1	1	1	1	1	1	1
19	20	21	22	23	24	25	26	27	28	29	30	31	32	33	34	35	36
1	1	1	1	1	1	1	1	1	1	1	1	1	1	1	1	2	2
37	38	39	40	41	42	43	44	45	46	47	48	49	50	51	52	53	54
2	2	2	2	2	2	2	2	2	2	2	2	2	2	3	3	3	3
55	56	57	58	59	60	61	62	63	64	65	66	67	68	69	70	71	72
3	3	3	3	3	3	3	3	3	3	3	3	3	4	4	4	4	4
73	74	75	76	77	78	79	80	81	82	83	84	85	86	87	88	89	90
4	4	4	4	4	4	4	4	4	4	4	4	5	5	5	5	5	5
91	92	93	94	95	96	97	98	99	100	101	102	103	104	105	106	107	108
5	5	5	6	6	6	6	6	6	6	6	6	6	7	7	7	7	7
109	110	111	112	113	114	115	116	117	118	119	120	121	122	123	124	125	126
7	7	7	7	7	7	7	7	7	7	7	7	7	7	7	7	7	7
127	128	129	130	131	132	133	134	135									
8	8	8	8	8	8	8	8	8									

```
str(DataRaw)
Cls_prior=DataRaw$Cls
Cls_prior
# if you want the full unsupervised way

ind=which(colnames(DataRaw)!=&quot;Cls&quot;)
Data=as.matrix(DataRaw[,ind])
library(DatabionicSwarm)
projection = Pswarm(Data,
Cls = Cls_prior,
PlotIt = T,
Silent = T)

library(DatabionicSwarm)
library(GeneralizedUmatrix)
visualization = GeneratePswarmVisualization(Data = Data,
projection$ProjectedPoints,
projection$LC)

GeneralizedUmatrix::plotTopographicMap(visualization
$Umatrix,
visualization$Bestmatches)

library(DatabionicSwarm)
library(GeneralizedUmatrix)
Cls = DBSclustering(k = 8,
Data,
visualization$Bestmatches,
visualization$LC,
PlotIt = FALSE)
FCPS::ClusterCount(Cls)
GeneralizedUmatrix::plotTopographicMap(visualization
$Umatrix,
visualization$Bestmatches,
Cls)
FCPS::ClusterAccuracy(PriorCls = Cls_prior, Cls)

library(DataVisualizations) #(Thrun, 2021a)
Heatmap(as.matrix(dist(Data)),Cls = Cls)
Silhouetteplot(Data,Cls =Cls)
```

In this case, the accuracy was 100 percent, so the assignments were made without error. Other exemplary validation indexes, known as nonsupervised indexes, used for evaluating the quality of the clustering are the silhouette index (Kaufman and Rousseeuw, 2005), the Dunn index (Dunn, 1974), and the

Davies Bouldin index (Davies and Bouldin, 1979). Furthermore, visualization techniques like the heatmap (Wilkinson and Friendly, 2009) or the Silhouette plot can be used. Some of these are called in the last part of the previous script. For example, in the silhouette plot, the ideal number of clusters is displayed as separated silhouettes within a range of values that go from -1 to $+1$, where $+1$ indicates that the samples are correctly assigned to a cluster, 0 shows that the observations are very close to the decision limit between two neighboring clusters, and negative values indicate that the samples might have been assigned to the wrong cluster. The result of the silhouette plot (Figure 27, left image) clearly marks the presence of eight clusters as the optimal number, with no negative observations and no zero values. Therefore, we can sustain with confidence the existence of eight groups.

Another graphical representation for visualizing high-dimensional data is the heatmap. A heatmap visualizes the distances ordered by the clustering through variations in color; this display simultaneously reveals row and column hierarchical cluster structure in a data matrix (Wilkinson and Friendly, 2009). In this example, the heatmap (Figure 27, right image) confirmed the DBS clustering of eight separated clusters; it also showed that this dataset was defined by discontinuities with small intracluster distances and large intercluster distances. Hence, the obsidian set is a high-dimensional dataset with natural clusters that are specified by the values represented in the two intervals of the spectra.

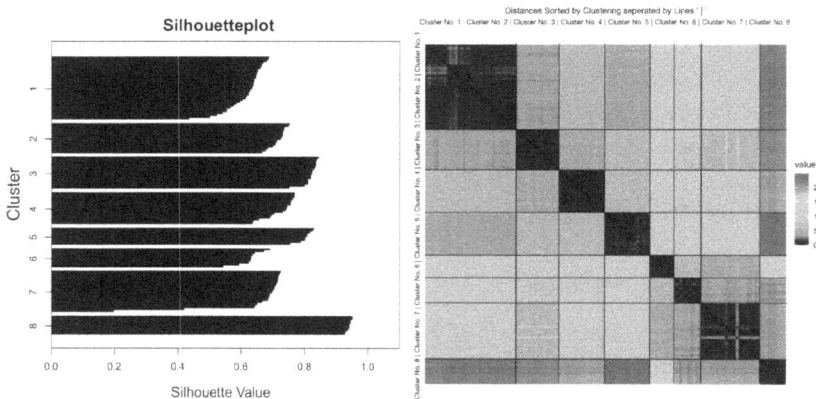

Figure 27 (Left) Silhouette plot of the obsidian dataset indicates a cluster structure. (Right) Heatmap of the obsidian dataset showing the existence of eight groups, where the intracluster distances are distinctively smaller than the intercluster distances. Color version available at www.cambridge.org/argote_machine-learning

The clustering is also evaluated by means of a contingency table, whose rows are the source groups and whose columns are the results of the clustering. This contingency table is computed by the last part of the script of the projection-based clustering algorithm, copied below:

```
## With the following script, the contingency table is calcu-
lated
Cls_prior=DataRaw$Cls
Cls_prior
ind=which(colnames(DataRaw)!="Cls")
Data=as.matrix(DataRaw[,ind])
ContingencyTableSummary=function(RowCls, ColCls)
{
# contingency table of two Cls
# INPUT
# RowCls,bCls vector of class identifiers (i.e., integers or
NaN's) of the same length
# OUTPUT list with these elements:
# cTab cTab(i,j) contains the count of all instances where the i-
th class in RowCls
#equals the j-th class inColCls
# rowID the different classes in RowCls, corresponding to the
rows of cTab
# colID the different classes inColCls, corresponding to the
columns of cTab
# RowClassCount, RowClassPercentages instance count and per-
centages of classes in
#RowCls sorted according rowID
# ColClassCount, ColClassPercentages instance count and per-
centages of classes #inColCls sorted according colID
RowID = length(unique(RowCls))
ColID = length(unique(ColCls))
Ctable = table(RowCls, ColCls)
AllinTab = sum(Ctable)
ColumnSum = colSums(Ctable)
ColPercentage = round(ColumnSum/AllinTab * 100, 2)
RowSum = rowSums(Ctable)
RowPercentage = round(RowSum/AllinTab * 100, 2)
Rows <- rbind(round(Ctable), ColumnSum, ColPercentage)
Xtable <- cbind(Rows, c(RowSum, AllinTab, 0), c(RowPercentage,
0, 100))
colnames(Xtable) = c(1:ColID, "RowSum", "RowPercentage")
return(Xtable)
```

```
}
Table=ContingencyTableSummary(Cls_prior,Cls)
Table
```

Table 14 shows that all the elements of the main diagonal are well classified and that there are no misclassified observations. According to the overall results, it was shown that the analyzed obsidian samples had a clear cluster structure linked to their geological origin.

In sum, DBS is a nonlinear projection that displays the structure of the high-dimensional data into a low-dimensional space, preserving the cluster structure of the data. This model exploits the concepts of self-organization and emergence, game theory and swarm intelligence (Thrun, 2018; Thrun and Ultsch, 2021). Pswarm does not require any input parameters other than the dataset of interest and is able to adapt itself to structures of high-dimensional data such as natural clusters characterized by distance and/or density-based structures in the data space. The result of the clustering consists of a 3D landscape with hypsometric tints, where observations with similar characteristics are represented as valleys while differences are represented as mountain ranges. In addition, the procedure can detect outliers that are represented as volcanoes on the 3D display and can be interactively marked on the display after the automated grouping process.

Another advantage of the method is that it is not necessary to have *a priori* knowledge of the classes to which the observations belong; the number of clusters and the cluster structure can be estimated by counting the valleys in the 3D topographic map and from the silhouette plot. Unlike other clustering algorithms, Pswarm does not impose any type of geometric structure in the formation of clusters, and the user does not need to specify any parameters. The results are evaluated using supervised and unsupervised validation indexes and visualization techniques, as well as a contingency table, confirming the goodness of the model to detect natural groups in the dataset. The example confirmed that the method proposed here is suitable for handling unbiased quantitative spectral analysis of archaeological materials.

4.3 Exercise 2: Human-in-the-Loop Cluster Analysis

The practical case study below serves as motivation to outline an alternative to investigate the detection and recognition of cluster structures using a human-in-the-loop. In this context, higher-level structures are detected in the data by enabling recognition of structures by the human-in-the-loop (HIL) at critical decision points. The authors thus follow the reasoning of Holzinger (2018) in

Table 14 Contingency table computed from the data of the obsidian sources.

	1	2	3	4	5	6	7	8	Rowsum	RowPercentage
1	34.00	0.00	0.00	0.00	0.00	0.00	0.00	0.00	34	25.19
2	0.00	16.00	0.00	0.00	0.00	0.00	0.00	0.00	16	11.85
3	0.00	0.00	17.00	0.00	0.00	0.00	0.00	0.00	17	12.59
4	0.00	0.00	0.00	17.00	0.00	0.00	0.00	0.00	17	12.59
5	0.00	0.00	0.00	0.00	9.00	0.00	0.00	0.00	9	6.67
6	0.00	0.00	0.00	0.00	0.00	10.00	0.00	0.00	10	7.41
7	0.00	0.00	0.00	0.00	0.00	0.00	22.00	0.00	22	16.3
8	0.00	0.00	0.00	0.00	0.00	0.00	0.00	10.00	10	7.41
Columsum	34.00	16.00	17.00	17.00	9.00	8.00	24.00	10.00	135	0.00
ColPercentage	25.2	11.9	12.6	12.6	6.7	5.93	17.8	7.41	0	100

that "the integration of an HIL's knowledge, intuition, and experience can sometimes be indispensable, and the interaction of an HIL with the data can significantly improve the overall ML pipeline. Such interactive ML uses the HIL to make possible what neither a human nor a computer could do alone (Holzinger, 2018). The HIL is an agent that interacts with algorithms, allowing the algorithms to optimize their learning behavior (Holzinger et al., 2019). This perspective fundamentally integrates humans into an algorithmic loop with the goal of opportunistically and repeatedly using human knowledge and skills to improve the quality of ML systems (Holzinger et al., 2019; see also Mac Aodha et al., 2014; Zanzotto, 2019).

An HIL is usually necessary because automatic detection pipelines (see example in Wiwie et al., 2015) have pitfalls and challenges that are systematically highlighted by Thrun (2021b). The work shows that parameter optimization on datasets without distance-based structures, algorithm selection using unsupervised quality measures on biomedical data, and benchmarking of detection algorithms with first-order statistics or box plots or a small number of repetitions of identical algorithm calls are biased and often not recommended (Thrun, 2021b). Hence, an alternative is proposed in Thrun et al. (2020; 2021a). Unlike typical approaches, the HIL is not overwhelmed with extensive parameter settings or evaluation of many complex quality measures (see examples in Choo et al., 2013; Müller et al., 2008; Yang et al., 2019; Zhang et al., 2002).

The coexistence of nonlinear projection methods and automatic detection of structures in projection-based clustering allows – apart from estimating whether there is a tendency for separable high-dimensional structures – estimating the number of partitions in the data as well as the correct choice of only one Boolean parameter for projection-based clustering. The HIL extension of the projection-based clustering incorporates user decisions in the detection process to visually discriminate structures. HIL-projection-based clustering is an open-source method that integrates the HIL at critical decision points through an interactive topographic map to detect separable structures.

Comparable interactive approaches fall into the category of visual analytics, which use visualizations to assist in manually searching for partitions in various types of datasets or to check the results of detection algorithms (e.g., Cavallo and Demiralp, 2018; Jeong et al., 2009; Kwon et al., 2017; Müller et al., 2008; Rasmussen and Karypis, 2004). However, the Johnson-Lindenstrauss lemma states that two-dimensional similarities in a scatter plot do not necessarily represent high-dimensional structures (Dasgupta and Gupta, 2003; Johnson and Lindenstrauss, 1984). In praxis, projections of several datasets with distance- and density-based structures show a misleading interpretation of the underlying structures and unsupervised quality measures for dimensionality reduction are biased toward assumed underlying structures (Thrun et al., 2023).

The HIL-projection-based clustering is proposed in a toroidal 2.5D representation, where a zoom out is preferred instead of other possible alternatives such as the four-tile of the toroidal projection. Toroid means that the boundaries of the topographic map are cyclically connected (Ultsch, 1999), which avoids problems of projections at the edges and therefore edge effects. In the four-tile representation, each projection point and structure would be represented four times. The HIL detects the number of partitions as the number of valleys. After the automatic detection phase, the HIL interactively rectifies the result on the topographic map via its own detection of high-dimensional structures. Using the 2.5D representation of the topographic map avoids the drawbacks of 3D representations and eliminates the challenge of the Johnson–Lindenstrauss lemma. The performance of HIL-projection-based clustering outperforms other accessible methods both qualitatively and quantitatively (Thrun et al., 2020, 2021b).

The GUI for the HIL-projection-based clustering is called with the following script:

```
rm(list=ls())
library(FCPS)
library(ProjectionBasedClustering) ## (Thrun et al., 2020)
data("Chainlink",package = "FCPS")
str(Chainlink)
Data <- Chainlink$Data
str(Data)
V= IPBC (Data)
#with prior classification
Cls=Chainlink$Cls
V=IPBC(Data,Cls)
```

The interfaces of the HIL-projection-based clustering algorithm (Thrun et al., 2021a) are presented in Figures 28 ('Projection' menu) and 29 ('Clustering' menu), in which every parameter setting is listed and numbered. In the 'Projection' menu, after loading the Chainlink dataset in listing (1), selecting the NeRV projection in (2), and clicking on the button in (4), the topographic map shown in (10) is obtained. The user can select another projection method in (2) and click in (4) to visualize a new topographic map. The user can automatically cluster the data with (15) by setting the number of clusters in (13) as the number of valleys. If the automatic clustering does not overlap with the valleys, the critical parameter for the clustering in (14) can be changed and

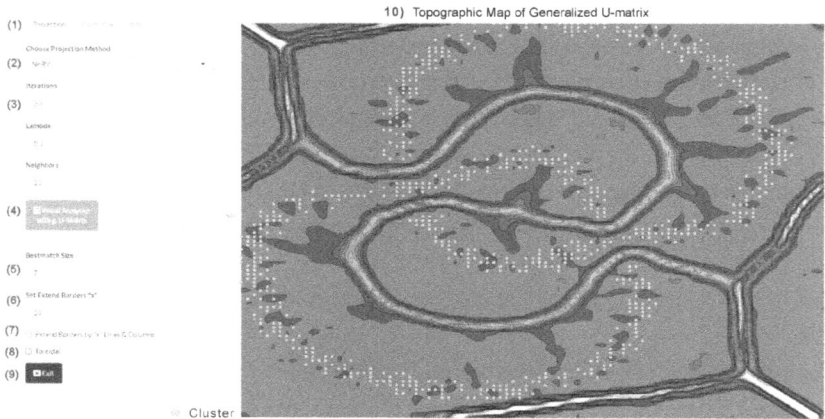

Figure 28 Screenshot of the interface of the "Projection" menu of the IPBC method. Color version available at www.cambridge.org/argote_machine-learning

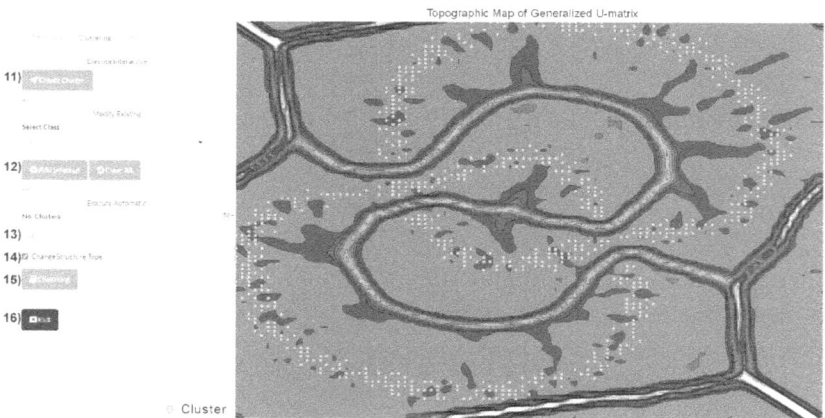

Figure 29 Screenshot of the interface of the "clustering" menu of the IPBC method. Color version available at www.cambridge.org/argote_machine-learning

a second trial of clustering can be performed. The user can frame points with the mouse either in the borderless view or the toroidal view (8); with "Add Cluster" in (12), a new cluster is assigned. Be aware that the assumption is that the data had been preprocessed properly and that the parameter of the selected projection method had been chosen wisely. If parameter setting of a projection method seems to be a challenge, the parameter-free projection method Pswarm can be selected.

5 Final Comments

One of the problems faced by archaeologists is that of data classification, which is defined by a series of attributes. Both clustering and classification have in common creating a model capable of recognizing instances according to their attributes by assigning them to different classes or groups. Both are complex tasks in archaeological data, since they involve the choice between many methods, the transformation and diagnosis of data, the selection of parameters and different metrics. Traditional unsupervised and supervised methods perform poorly at uncovering the underlying group structure in the data because they lack a formal statistical model. This Element compares the performance of different supervised and unsupervised classification methods that improve the outcome in clustering and classification of archaeological data.

If the data is high-dimensional, as in the case of pXRF spectral quantification, pre-processing is an essential part of comprehensive analysis to improve data quality. For example, spectra can show noise, displacements, or overlap between elements. To correct these problems in the related examples, a simple method for peak alignment was proposed using the hierarchical Cluster-based Peak Alignment (CluPA) algorithm. CluPa takes care of bringing all the peaks to the same origin, showing that the peaks from different origins are not misaligned. To eliminate unwanted interference in the spectra, model-based pre-processing techniques allow to quantify and separate different types of physical and chemical variations in the spectra. The recommended filtering was the use of a combination of the Savitzky–Golay and Extended Multiplicative Scatter Correction (EMSC) algorithms, which are filters that allow the preservation of the main characteristics of the function such as width and height of the spectral peaks.

The diagnosis of the data was another point to consider since extreme values can seriously distort the behavior of statistical estimators. To detect outliers, a robust estimator was proposed by replacing the covariance matrix obtained with the classical method with the covariance matrix obtained with the Minimum Covariance Determinant (MCD) robust method. On the other hand, a feature selection procedure is essential to separate variability related to relevant information from non-relevant information. In the high-dimensional (spectral) data scenarios, two variable selection methods were proposed. One approach based on Bayesian models and the other based on Partial Least Squares Interval (iPLS). The Bayesian approach allows for the selection of relevant variables and clustering simultaneously; at the same time is able to automatically determine the memberships of instances to their respective groups and to determine the optimal number of groups in the data. On the other hand, the iPLS method allows the selection of specific spectral regions for further analysis.

Continuing with the spectral analysis, Databionic swarm (DBS) algorithm propose an approach in which dimensionality reduction methods coexist with clustering algorithms, using a swarm-based AI system. The resulting groups define different generating processes related to the chemical composition of each group, which can be visualized with a topographic map of high-dimensional structures. The central problem in clustering is the correct estimation of the number of clusters; this is also addressed by the topographic map, which evaluates the optimal number of clusters. This is important because most clustering algorithms can detect clusters even if the distribution of the data is random. Unlike traditional methods, DBS is not going to force a cluster if there really are not natural clusters in the data.

When dealing with compositional data, first, it is important to consider that geometric space is different from real Euclidean space. In general, standard statistical methods are designed to work with classical Euclidean geometry (R^D) or unconstrained p-dimensional spaces, so it is advisable to use a suitable transformation for compositional data whose geometric space is the simplex (S^D). This is a bounded space with a constant-sum constraint, in which Aitchison's geometry is applied; so only by working with the ratios between the parts do the problems of the constant-sum constraint disappear. There are many clustering methods in the literature without any specific method being uniformly better. To select the most suitable one, it is important to understand the intrinsic nature of the data and the strengths and weaknesses of the different algorithms.

To classify compositional data, an alternative is to model the data using components of the Gaussian mixture distributions, which assumes that the sample to be classified is divided into G groups or components. The estimation of parameters for each group is estimated by the maximum likelihood method, allowing us to estimate the probability that each of the instances has of belonging to one of the classes. To find the right number of groups, information criteria are considered, which are statistical criteria for evaluating models in terms of their *posteriori* probabilities. These statistics allow you to select from among the competing models and determine the optimal from a finite family of models.

Because of the excellent performance in partitioning the experimental datasets, the suggested algorithms and methodologies applied in this Element have proved to work well in the difficult task of clustering when the number of groups is unknown. The advantage of the proposed methods is that they allow the creation of a model for the determination of groups of well-defined characteristics that allows the optimization of the classification of the data. It is worth mentioning that the software for the implementation of the proposed methods is freely accessible, which allows the implementation to be established in an easy and simple way. We believe that these methods can be of great use to the archaeological community, as well as being applicable to a large number of cases beyond those described here. We encourage readers to practice with the proposed techniques and try new ways to solve their research problems.

Abbreviations

alr	Additive Log-ratio
BIC	Bayesian Information Criterion
CA	Cluster Analysis
CAIS	Center for Applied Isotope Studies
CEM	Classification Expectation Maximization
clr	Centered Log-ratio
CluPA	Cluster-based Peak Alignment
CRM	Certified Reference Material
DA	Discriminant Analysis
DBS	Databionic Swarm
DR	Dimensionality Reduction
EM	Expectation-Maximization
EMSC	Extended Multiplicative Signal Correction
FAST-MCD	Fast Minimum Covariance Determinant
FCPS	Fundamental Clustering Problems Suite
FT-IR	Fourier Transform Infrared Spectroscopy
HIL	Human-in-the-Loop
ICP–MS	Inductively Coupled Plasma–Mass Spectrometry
ICL	Integrated Complete Likelihood
ilr	Isometric Log-ratio
iPLS	Interval Partial Least Squares
LDA	Linear Discriminant Analysis
LV	Latent Variables
MAP	Maximum a Posteriori Probability
MCD	Minimum Covariance Determinant
MD	Mahalanobis Distance
MD-plot	Mirrored Density plot
ML	Maximization of Log-likelihood
n	Number of observations or samples
NAA	Neutron Activation Analysis
NEC	Normalized Entropy Criterion
NPE	Neighborhood Proportion Error
ODi	Orthogonal Distance
p	Number of variables or components
PBC	Projection-based clustering
PCA	Principal Component Analysis

PC	Principal Component
PDE	Pareto Density Estimation
PDF	Probability Density Function
PLS	Partial Least Squares
PP	Projection Pursuit
pXRF	Portable X Ray Fluorescence
QDA	Quadratic Discriminant Analysis
RMSECV	Root Mean Squared Error of Cross-Validation
RMSEP	Root Mean Square Error of Prediction
ROBPCA	Robust Principal Component Analysis
SBP	Sequential Binary Partition
SDi	Score Distance
SEM	Stochastic Expectation-Maximization
SG	Savitzky–Golay filter
SNE	Stochastic Neighbor Embedding
SVD	Singular Values Decomposition
XRF	X Ray Fluorescence

References

Abascal, R. (1974). *Análisis por Activación de Neutrones: Una Aportación para la Arqueología Moderna*, B.A. Thesis in Archaeology, Escuela Nacional de Antropologia e Historia, Mexico City, Mexico.

Aitchison, J. (1986). *The Statistical Analysis of Compositional Data*. London: Chapman & Hall.

Argote-Espino, D. L., Solé, J., Sterpone, O. & López García, P. (2010). Análisis composicional de seis yacimientos de obsidiana del centro de México y su clasificación con DBSCAN. *Arqueología*, **43**, 197–215.

Argote-Espino, D. L., Solé, J., López-García, P. & Sterpone, O. (2012). Obsidian sub-source identification in the Sierra de Pachuca and Otumba volcanic regions, Central Mexico, by ICP-MS and DBSCAN statistical analysis. *Geoarchaeology*, **27**, 48–62.

Aubert, A. H., Thrun, M. C., Breuer, L. & Ultsch, A. (2016). Knowledge discovery from high-frequency stream nitrate concentrations: Hydrology and biology contributions. *Scientific Reports*, **6**, 31536.

Biernacki, C., Marbac, M. & Vandewalle, V. (2021). Gaussian-based visualization of gaussian and non-gaussian-based clustering. *Journal of Classification*, **38**(1), 129–157.

Brambila, R. (1988). Los estudios de la cerámica Anaranjada Delgada: ensayo bibliográfico. In M. C. Serra Puche and C. Navarrete, coords., *Ensayos de Alfarería prehispánica e histórica de Mesoamérica. Homenaje a Eduardo Noguera Auza*. México: Instituto de Investigaciones Antropológicas and Universidad Nacional Autónoma de México, pp. 221–247.

Brinkmann, L., Stier, Q. & Thrun, M. C. (2023). Computing sensitive color transitions for the identification of two-dimensional structures. In *Proceedings of Data Science, Statistics & Visualisation and the European Conference on Data Analysis*. Antwerp: University of Antwerp, DSSV-ECDA, pp. 57.

Callaghan, M. G., Pierce, D. E., Kovacevich, B. & Glascock, M. D. (2017). Chemical paste characterization of late middle Preclassic-period ceramics from Holtun, Guatemala and its implications for production and exchange. *Journal of Archaeological Science: Reports*, **12**, 334–345.

Carr, S. (2015). *Geochemical Characterization of Obsidian Subsources in Highland Guatemala*, B.A. Thesis, Pennsylvania State University, US.

Cavallo, M. & Demiralp, Ç. (2018). Clustrophile 2: Guided visual clustering analysis. *IEEE Transactions on Visualization and Computer Graphics*, **25**(1), 267–276.

Choo, J., Lee, H., Liu, Z., Stasko, J. & Park, H. (2013). An interactive visual testbed system for dimension reduction and clustering of large-scale high-dimensional data. In *Proceedings of SPIE-IS and T Electronic Imaging – Visualization and Data Analysis 2013* [8654]. Burlingame, CA: The International Society for Optics and Photonics, 865402.

Cobean, R. H. (2002). *A World of Obsidian: The Mining and Trade of a Volcanic Glass in Ancient Mexico*. Mexico: Instituto Nacional de Antropologia e Historia and Pittsburgh University.

Cook de Leonard, C. (1953). Los popolocas de Puebla (ensayo de una identificación etnodemográfica e histórico arqueológica), Huastecos, Totonacos y sus vecinos. *Revista Mexicana de Estudios Antropológicos*, **13** (2–3), 423–445.

Dasgupta, S. & Gupta, A. (2003). An elementary proof of a theorem of Johnson and Lindenstrauss. *Random Structures & Algorithms*, **22**(1), 60–65.

Davies, D. L. & Bouldin, D. W. (1979). A cluster separation measure. *IEEE Transactions on Pattern Analysis and Machine Intelligence*, **1**(2), 224–227.

Dunn, J. C. (1974). Well-separated clusters and optimal fuzzy partitions. *Journal of Cybernetics*, **4**(1), 95–104.

Egozcue, J. J., Pawlowsky-Glahn, V., Mateu-Figueras, G. & Barceló-Vidal, C. (2003). Isometric logratio transformations for compositional data analysis. *Mathematical Geology*, **35**(3), 279–300.

Glascock, M. D., Braswell, G. E. & Cobean, R. H. (1998). A systematic approach to obsidian source characterization. In M. S. Shackley, ed., *Archaeological Obsidian Studies*. Vol 3 of *Advances in Archaeological and Museum Science*. Boston, MA: Springer, pp. 15–65.

Grün, B. (2019). Chapter 8: Model-based clustering. In S. Frühwirth-Schnatter, G. Celeux, and C. P. Robert, eds., *Handbook of Mixture Analysis*. Boca Raton, FL: Chapman and Hall/CRC Press, pp. 1–36.

Harbottle, G., Sayre, E. V. & Abascal R. (1976). *Neutron Activation Analysis of Thin Orange Pottery*. Upton, NY: Brookhaven National Lab.

Holzinger, A. (2018). From machine learning to explainable AI. *IEEE, Proceedings of the 2018 World Symposium on Digital Intelligence for Systems and Machines (DISA)*, 55–66.

Holzinger, A., Plass, M., Kickmeier-Rust, M. et al. (2019). Interactive machine learning: Experimental evidence for the human in the algorithmic loop. *Applied Intelligence*, **49**(7), 2401–2414.

Horikoshi, M., Tang, Y., Dickey, A. et al. (2023). *Package 'ggfortify' Version 0.4.16: Data Visualization Tools for Statistical Analysis Results*. https://cran.r-project.org/web/packages/ggfortify/ (Accessed: August 03).

Hubert, M., Rousseeuw, P. J. & Vanden Branden, K. (2005). ROBPCA: A new approach to robust principal component analysis. *Technometrics*, **47**(1), 64–79.

Hunt, A. M. W. & Speakman, R. J. (2015). Portable XRF analysis of archaeological sediments and ceramics. *Journal of Archaeological Science*, **53**, 626–638.

Jain, A. K. & Dubes, R. C. (1988). *Algorithms for Clustering Data*. Vol. 3. Englewood Cliffs, NJ: Prentice Hall College Division.

Jeong, D. H., Ziemkiewicz, C., Fisher, B., Ribarsky, W. & Chang, R. (2009). iPCA: An Interactive system for PCA-based visual analytics. *Computer Graphics Forum*, **28**(3), 767–774.

Johnson, W. B. & Lindenstrauss, J. (1984). Extensions of Lipschitz mappings into a Hilbert space. *Contemporary Mathematics*, **26**(1), 189–206.

Kaufman, L. & Rousseeuw, P. J. (2005). *Finding Groups in Data: An Introduction to Cluster Analysis*. Hoboken, NJ: Wiley-Interscience.

Kessler, D. (2019). *Introducing the MBC Procedure for Model-Based Clustering*. www.sas.com/content/dam/SAS/support/en/sas-global-forum-proceedings/2019/3016-2019.pdf (Accessed: December 09, 2020).

Kolb, C. (1973). Thin Orange Pottery at Teotihuacan. In W. Sanders, ed., *Miscellaneous Papers in Anthropology*, Vol. 8. Pennsylvania: Pennsylvania State University, pp. 309–377.

Kucheryavskiy, S. (2020). mdatools – R package for chemometrics. *Chemometrics and Intelligent Laboratory Systems*, **198**, 103937.

Kwon, B. C., Eysenbach, B., Verma, J. et al. (2017). Clustervision: Visual supervision of unsupervised clustering. *IEEE Transactions on Visualization and Computer Graphics*, **24**(1), 142–151.

Lebret, R., Lovleff, S., Langrognet, F. et al. (2015). Rmixmod: The R package of the model-based unsupervised, supervised, and semi-supervised classification Mixmod library. *Journal of Statistical Software*, **67**(6), 1–29.

Liland, K. H. & Indahl, U. G. (2020). Package "EMSC". *Extended Multiplicative Signal Correction*. https://cran.r-project.org/web/packages/EMSC/EMSC.pdf (Accessed: August 22, 2020).

Linné, S. (2003). *Mexican Highland Cultures: Archaeological Researches at Teotihuacan, Calpulalpan and Chalchicomula in 1934–1935*. Alabama: The University of Alabama Press.

López-García, P. A. & Argote D. L. (2023). Cluster analysis for the selection of potential discriminatory variables and the identification of subgroups in archaeometry. *Journal of Archaeological Science: Reports*, **49**, 104022.

López-García, P., Argote, D. L. & Beirnaert, C. (2019). Chemometric analysis of Mesoamerican obsidian sources. *Quaternary International*, **510**, 100–118.

López-García, P., Argote, D. L. & Thrun, M. C. (2020). Projection-based classification of chemical groups and provenance analysis of archaeological materials. *IEEE Access*, **8**, 152439–152451.

López Luján, L., Neff, H. & Sugiyama, S. (2000). The 9-Xi Vase: A Classic Thin Orange vessel found at Tenochtitlan. In D. Carrasco, L. Jones, and S. Sessions, coords., *Mesoamerica's Classic Heritage: From Teotihuacan to the Aztecs*. Boulder, CO: University Press of Colorado, pp. 219–249.

Lukas-Tooth, H. J. & Price, B. J. (1961). A mathematical method for the investigation of interelement effects in x-ray fluorescence analysis. *Metallurgia*, **64**(2), 149–152.

Mac Aodha, O., Stathopoulos, V., Brostow, G. J., et al. (2014). Putting the scientist in the loop–accelerating scientific progress with interactive machine learning. In *Proceedings of the 2014 22nd International Conference on Pattern Recognition*. Stockholm: IEEE, pp. 9–17.

Maechler, M., Rousseeuw, P., Struyf, A., Hubert, M. & Hornik, K. (2022). Cluster: Cluster Analysis Basics and Extensions. *R Package Version 2.1.4*. https://CRAN.R-project.org/package=cluster (Accessed: August 03, 2023).

Minc, L. D., Sherman, R. J., Elson, C., et al. (2016). Ceramic provenance and the regional organization of pottery production during the later Formative periods in the Valley of Oaxaca, Mexico: Results of trace-element and mineralogical analyses. *Journal of Archaeological Science: Reports*, **8**, 28–46.

Müller, F., 1978. *La Cerámica del Centro Ceremonial de Teotihuacán*. México: Instituto Nacional de Antropología e Historia.

Müller, E., Assent, I., Krieger, R., Jansen, T. & Seidl, T. (2008). Morpheus: Interactive exploration of subspace clustering. *Proceedings of the 14th ACM SIGKDD International Conference on Knowledge Discovery and Data Mining*. Las Vegas, NV: ACM, pp. 1089–1092.

Murtagh, F. (2004). On ultrametricity, data coding, and computation. *Journal of Classification*, **21**(2), 167–184.

Nance, R. D., Mille, B. V., Keppie, J. D., Murphy, J. B. & Dostal, J. (2006). Acatlán Complex, southern Mexico: Record spanning the assembly and breakup of Pangea. *Geology*, **34**(10), 857–860.

Parent, S. É., Parent, L. E., Rozanne, D. E., Hernandes, A. & Natale, W. (2012). Chapter 4: Nutrient balance as paradigm of soil and plant chemometrics. In R. Nuhu Issaka, ed., *Soil Fertility*. London: InTechOpen, pp. 83–114.

Partovi Nia, V. & Davison, A. C. (2012). High-dimensional bayesian clustering with variable selection: The R package bclust. *Journal of Statistical Software*, **47**(5), 1–22.

Partovi Nia, V. & Davison, A. C. (2015). *Package "bclust": Bayesian Hierarchical Clustering Using Spike and Slab Models*. https://cran.micro

soft.com/snapshot/2017-07-05/web/packages/bclust/bclust.pdf   (Accessed: January 15, 2020).

Pawlowsky-Glahn, V., Egozcue, J. (2011). Exploring compositional data with the CoDa-dendrogram. *Austrian Journal of Statistics*, **40**(1–2), 103–113.

R Development Core Team (2011). *R: A language and environment for statistical computing*. Vienna: R Foundation for Statistical Computing. www.R-project.org (Accessed: November 24, 2016).

Rasmussen, M. & Karypis, G. (2004). Gcluto: An interactive clustering, visualization, and analysis system. CSE/UMN Technical Report: TR# 04–021, Department of Computer Science & Engineering, University of Minnesota, Minnesota, USA. https://hdl.handle.net/11299/215615 (Accessed: November 24, 2016).

Rattray, E. C. (1979). La cerámica de Teotihuacan: relaciones externas y cronología. *Anales de Antropología*, **16**, 51–70.

Rattray, E. C. (2001). *Teotihuacan: Ceramics, Chronology and Cultural Trends*. Mexico: Instituto Nacional de Antropología e Historia – University of Pittsburgh.

Rattray, E. C. & Harbottle, G. (1992). Neutron Activation Analysis and numerical taxonomy of Thin Orange ceramics from the manufacturing sites of Rio Carnero, Puebla, Mexico. In H. Neff, ed., *Chemical Characterization of Ceramic Pastes in Archaeology*. Madison, WI: Prehistory Press, pp. 221–231.

Rousseeuw, P. J. & van Zomeren, B. C. (1990). Unmasking multivariate outliers and leverage points. *Journal of the American Statistical Association*, **85**(411), 633–639.

Rowe, H., Hughes, N. & Robinson, K. (2012). The quantification and application of handheld energy-dispersive X-ray fluorescence (ED-XRF) in Mudrock Chemostratigraphy and Geochemistry. *Chemical Geology*, **324–325**, 122–131.

Ruvalcaba-Sil, J. L., Ontalba Salamanca, M. A., Manzanilla, L. et al. (1999). Characterization of pre-Hispanic pottery from Teotihuacan, Mexico, by a combined PIXE-RBS and XRD analysis. *Nuclear Instruments and Methods in Physics Research B: Beam Interactions with Materials and Atoms*, **150**(1–4), 591–596.

Shepard, A. O. (1946). Technological features of Thin Orange Ware. In A. Kidder, J. Jenning, and E. Shook, eds., *Excavations at Kaminaljuyu*. Publication 561. Washington, DC: Carnegie Institution of Washington, pp. 198–201.

Sotomayor, A. & Castillo, N. (1963). *Estudio Petrográfico de la Cerámica "Anaranjada Delgada"*. Publicaciones del Departamento de Prehistoria no. 12. México: Instituto Nacional de Antropología e Historia.

Stevens, A. & Ramirez-Lopez, L. (2015). *Package 'prospectr': Miscellaneous functions for processing and sample selection of Vis-NIR Diffuse Reflectance Data*. https://cran.rproject.org/web/packages/prospectr/prospectr.pdf (Accessed: April 24, 2017).

Stevens, A., Ramirez-Lopez, L. & Hans, G. (2022). *Package "prospectr": Miscellaneous Functions for Processing and Sample Selection of Spectroscopic Data Version 0.2.4*. https://mran.microsoft.com/web/packages/prospectr/prospectr.pdf (Accessed: April 17, 2022).

Stoner, W. D. (2016). The analytical nexus of ceramic paste composition studies: A comparison of NAA, LA-ICP-MS, and petrography in the prehispanic Basin of Mexico. *Journal of Archaeological Science*, **76**, 31–47.

Thrun, M. C. (2018). *Projection Based Clustering through Self-Organization and Swarm Intelligence: Combining Cluster Analysis with the Visualization of High-Dimensional Data* (Ultsch, A. & Hüllermeier, E., eds.). Heidelberg: Springer Vieweg.

Thrun, M. C. (2021a). The exploitation of distance distributions for clustering. *International Journal of Computational Intelligence and Applications*, **20**(3), 2150016.

Thrun, M. C. (2021b). Distance-based clustering challenges for unbiased benchmarking studies. *Scientific Reports*, **11**, 18988.

Thrun, M. C. (2022). Exploiting distance-based structures in data using an explainable AI for stock picking. *Information*, **13**(2), 51

Thrun, M. C. & Ultsch, A. (2021). Swarm intelligence for self-organized clustering. *Artificial Intelligence*, **290**, 103237.

Thrun, M. C., Lerch, F., Lötsch, J. & Ultsch, A. (2016). Visualization and 3D printing of multivariate data of biomarkers. In V. Skala, ed., 24th *International Conference in Central Europe on Computer Graphics, Visualization and Computer Vision (WSCG)*. Plzen, Czech Republic, pp. 7–16.

Thrun, M. C., Pape, F. & Ultsch, A. (2020). Interactive machine learning tool for clustering in visual analytics. In *7th IEEE International Conference on Data Science and Advanced Analytics (DSAA)*. Sydney: IEEE, pp. 672–680.

Thrun, M. C., Pape, F. & Ultsch, A. (2021a). Conventional displays of structures in data compared with interactive projection-based clustering (IPBC). *International Journal of Data Science and Analytics*, **12**(3), 249–271.

Thrun, M. C., Ultsch, A., & Breuer, L. (2021b). Explainable AI framework for multivariate hydrochemical time series. *Machine Learning and Knowledge Extraction*, **3**(1), 170–205.

Thrun, M. C., Mack, E., Neubauer, A. et al. (2022). A bioinformatics view on acute myeloid leukemia surface molecules by combined Bayesian and ABC analysis. *Bioengineering*, **9**(11), 642.

Thrun, M. C., Märte, J. & Stier, Q. (2023). Analyzing quality measurements for dimensionality reduction. *Machine Learning and Knowledge Extraction*, **5** (3), 1076–1118.

Todorov, V. (2020). *Package 'rrcov': Scalable Robust Estimators with High Breakdown Point*. https://cran.r-project.org/web/packages/rrcov/rrcov.pdf (Accessed: September 24).

Todorov, V. & Filzmoser, P. (2009). An object-oriented framework for robust multivariate analysis. *Journal of Statistical Software*, **32**(3), 1–47.

Ultsch, A. (1999). Data mining and knowledge discovery with emergent self-organizing feature maps for multivariate time series. In E. Oja and S. Kaski, eds., *Kohonen Maps*. Amsterdam: Elsevier Science B.V., pp. 33–46.

Ultsch, A. (2003). *U*-matrix: A tool to Visualize Clusters in High Dimensional Data*. Technical report 36, Department of Mathematics and Computer Science, University of Marburg, Germany.

Ultsch, A. & Siemon, H. P. (1990). Kohonen's self-organizing feature maps for exploratory data analysis. In *Proceedings of the International Neural Network Conference*. Paris: Kluwer Academic Press, pp. 305–308.

van den Boogaart, K. G. & Tolosana-Delgado, R. (2013). *Analyzing Compositional Data with R*. Heidelberg: Springer-Verlag.

van den Boogaart, K. G., Tolosana-Delgado, R. & Bren, M. (2023). *Package "compositions" versión 2.0–6: Compositional Data Analysis*. https://cran.r-project.org/web/packages/compositions/ (Accessed: August 03).

Walesiak, M. & Dudek, A. (2020). *Package "clusterSim": Searching for Optimal Clustering Procedure for a Data Set*. https://cran.r-project.org/web/packages/clusterSim/clusterSim.pdf (Accessed: November 15).

Wilkinson, L. & Friendly, M. (2009). The history of the cluster heat map. *The American Statistician*, **63**(2), 179–184.

Wiwie, C., Baumbach, J. & Röttger, R. (2015). Comparing the performance of biomedical clustering methods. *Nature Methods*, **12**, 1033–1038.

Yang, Y., Kandogan, E., Li, Y., Sen, P. & Lasecki, W. S. (2019). A study on interaction in human-in-the-loop machine learning for text analytics. In *Proceedings of the ACM IUI Workshops '19*. Los Angeles, CA: ACM, pp. 1–7.

Zanzotto, F. M. (2019). Human-in-the-loop artificial intelligence. *Journal of Artificial Intelligence Research*, **64**, 243–252.

Zhang, L., Tang, C., Shi, Y. et al. (2002). VizCluster: An interactive visualization approach to cluster analysis and its application on microarray data. In *Proceedings of the 2002 SIAM International Conference on Data Mining (SDM)*. Philadelphia, PA: Society for Industrial and Applied Mathematics, pp. 19–40.

Cambridge Elements ☰

Current Archaeological Tools and Techniques

Series Editors

Hans Barnard
Cotsen Institute of Archaeology

Hans Barnard was associate adjunct professor in the Department of Near Eastern Languages and Cultures as well as associate researcher at the Cotsen Institute of Archaeology, both at the University of California, Los Angeles. He currently works at the Roman site of Industria in northern Italy and previously participated in archaeological projects in Armenia, Chile, Egypt, Ethiopia, Italy, Iceland, Panama, Peru, Sudan, Syria, Tunisia, and Yemen.

Willeke Wendrich
Polytechnic University of Turin

Willeke Wendrich is Professor of Cultural Heritage and Digital Humanities at the *Politecnico di Torino* (Turin, Italy). Until 2023 she was Professor of Egyptian Archaeology and Digital Humanities at the University of California, Los Angeles, and the first holder of the Joan Silsbee Chair in African Cultural Archaeology. Between 2015 and 2023 she was Director of the Cotsen Institute of Archaeology, with which she remains affiliated. She managed archaeological projects in Egypt, Ethiopia, Italy, and Yemen, and is on the board of the International Association of Egyptologists, *Museo Egizio* (Turin, Italy), the Institute for Field Research, and the online UCLA Encyclopedia of Egyptology.

About the Series

Cambridge University Press and the Cotsen Institute of Archaeology at UCLA collaborate on this series of Elements, which aims to facilitate deployment of specific techniques by archaeologists in the field and in the laboratory. It provides readers with a basic understanding of selected techniques, followed by clear instructions how to implement them, or how to collect samples to be analyzed by a third party, and how to approach interpretation of the results.

COTSEN INSTITUTE OF
ARCHAEOLOGY AT UCLA

Cambridge Elements ≡

Current Archaeological Tools and Techniques